The CRYSTAL ALMANAC

HARNESS YOUR CRYSTALS THROUGH THE YEAR

Gemma Petherbridge

GODSFIELD

Dedication

I dedicate this book to you, the reader of *The Crystal Almanac*. This book is the result of a life's passion. Compiling it allowed me to fall in love with crystals, nature and the seasons all over again. May it now do the same for you. *The Crystal Almanac* is also a love letter to Mother Nature herself. I hope it goes some way towards honouring the magic of the planet we call home.

First published in Great Britain in 2024 by Godsfield Press, an imprint of Octopus Publishing Group Ltd, Carmelite House, 50 Victoria Embankment, London EC4Y 0DZ www.octopusbooks.co.uk www.octopusbooksusa.com

An Hachette UK Company www.hachette.co.uk

ISBN 978-1-84181-562-6

A CIP catalogue record for this book is available from the British Library.

Printed and bound in China.

10 9 8 7 6 5 4 3 2 1

FSC
MIX
Paper | Supporting responsible forestry
FSC® C008047

Disclaimer

No medical claims are made for crystals in this book and information given is not intended to act as a substitute for medical treatment. Healing means bringing mind, body and spirit back into balance, it does not imply a cure.

Sourcing crystals responsibly

Remember that crystals are a gift from Mother Earth, so please respect her and source your crystals responsibly. Ethical sourcing means educating yourself about where your crystals have come from, how they have been mined and the supply chain they have come through. The more we ask these questions, the more the crystal industry will realize how important these factors are to their customers.

Staff credits

Commissioning Editor: Louisa Johnson
Art Director: Yasia Williams
Designer & Illustrator: Claire Huntley
Senior Editor: Leanne Bryan
Copyeditor: Laura Gladwin
Picture Researchers: Giulia Hetherington & Jen Veall
Assistant Production Manager: Allison Gonsalves

Picture credits

Contents

How This Book Will Help You

The purpose of *The Crystal Almanac* is to show you how to enhance your life by living in line with seasons. In it, I'll teach you how you can use some of nature's greatest tools – crystals – to help to achieve this. I will walk you through the year, from January to December, and show you how to follow the seasonal cycles to support all areas of your life.

I have been a crystal therapist for 19 years now, but it wasn't until I started hosting traditional celebrations known as Sabbaths and moon circles that I began to explore how the different seasonal energies work magically together, creating a dynamic force that guides us through life.

I came to realize that if we work in unison with them, these energies can help us to achieve our dreams more quickly. This wise, ancient process, which I outline in this book, encourages us to set aside time each year to do the important work we might otherwise overlook – most notably, to review our achievements and heal from any toxic patterns or negative thoughts that might be holding us back from future growth.

These days, we are taught to work at the same pace all year, without stopping. If this is you, I'm sure you have come up against unforeseen obstacles: times when nothing seemed to be going in your favour and moments when you just wanted to rest but couldn't. All of these are signs we are working against nature rather than with nature. *The Crystal Almanac* is designed to help you understand where you are out of alignment with nature. I explain the dominant energies of each month and recommend crystals and seasonal activities that will bring you back in line with the seasonal flow.

This book will teach you how to harness the seasonal energies to accelerate your progress towards your goals. These could be any goals, big or small, and they don't have to be work-related. Perhaps you're saving for a new home, or renovating the one you have. Maybe you're on a fertility journey, or want more time to read. Or perhaps your goal is simply to live peacefully by the seasons. Everyone is on their own path, and the seasons can help us all.

NOT STARTING IN JANUARY?

If you're starting to read this book part way through the year, don't worry. It's a good idea to first read the book in its entirety so you understand all the seasonal cycles. You can then read each month as it begins to remind you of the energies to come, and the crystals that you might like to have with you along the way.

Crystal History

This book encourages you to reconnect to the natural world and deepen your appreciation for crystals. Although the seasonal energies are unruly, crystals act as natural tools we can use in everyday life to amplify the good and help us overcome the bad. Humans have long seen them as a protective tool, and crystal therapy is one of the oldest healing modalities. Rocks and crystals made the first tools; flint enabled us to hunt and prepare food. Gemstones make up our homes and adorn our jewellery.

To hold a crystal is to connect to the past. Consider how the crystal in your hand could be a distant relative of the giant stones used to construct megalithic structures, like the famous stone circles at Stonehenge. The oldest of these constructions, which dates from around 9000 BC, is Göbekli Tepe in Turkey, which is positioned to align with the brightest star in the night sky, Sirius. Other similar sites all around the world align with the dates of the solstices and equinoxes, suggesting how crystals are intrinsically connected to seasonal cycles.

Throughout history, there are many religious or mythic stories featuring crystals. The earliest crystal mines date back to the middle Stone Age, when materials were used for cosmetic and ritual work. Later on, the ancient Egyptians were the first to record the properties of crystals, while the ancient Greeks inspired the crystal teachings we refer to today. Mythological stories tell us of crystal connections to the gods and goddesses. For instance, the Greek goddess Selene shares her name with the crystal Selenite.

Other peoples have also told stories of how crystals were created. For example, the Inuit and Beothuk people believed the flash in Labradorite was a result of the northern lights (*aurora borealis*) becoming trapped in the stone. Around the world, ancient cultures embraced crystals' metaphysical properties, from Mayan crystal skulls to the wives of Roman centurions gifting their husbands Red Jasper as protection, to stop the shedding of their blood.

Almanac History

Historically, almanacs were regularly printed documents that marked useful dates and information such as weather forecasts, lunar cycles, tide tables and crop planting dates for farmers. The earliest texts considered to be almanacs date all the way back to Babylonia, while the first use of the word 'almanac' goes back to the thirteenth century. Over time, almanacs became useful annual resources and regular bestsellers in the UK and USA.

Today, almanacs cover a broad range of topics and themes. *The Crystal Almanac* is an exploration of crystal energy and how it relates to the seasons, and honours the content offered in the original almanacs.

For me, an almanac is a glimpse into our past: an acknowledgement of how hard life used to be and how important the seasons were to our ancestors. But today we can also use them as a map, a blueprint back to nature. When we feel disconnected from the natural world, working with crystals and understanding the seasons opens us up to experience their life-affirming, grounding and calming energies again.

The Wheel of the Year

In the following pages, I highlight the traditional seasonal celebrations known as Sabbaths. These festivals date all the way back to neolithic times, when they marked the summer and winter solstices and the two equinoxes. The Anglo Saxons carried on these celebrations, using them to mark the turning points of the year and as opportunities to pray to their gods and goddesses for protection and a prosperous, fertile year.

In western and central Europe, the Celtic tribes of the Bronze Age celebrated Samhain, Imbolc, Beltane and Lughnasadh. These are known as the cross-quarter fire festivals, and marked the peak of a season, when its energies are at their most potent. Even though each one was a fire festival, it also honoured that season's element and acknowledged significant moments in the agricultural and hunting year, offering communities a chance to celebrate, feast, dance, carry out rituals, give thanks and enjoy life before work started again.

However, with the rise of the Roman Empire and, later, Christianity, most people stopped honouring nature and our disconnection from the natural world slowly began. Fast forward to 1853, when Jacob Grimm (the famous collector of fairy tales) first suggested that all eight festivals (in other words, the solstices, equinoxes and the Celtic festivals) should be combined together. Gerald Gardner, the founder of modern-day Wicca (a religious movement based on the teachings of ancient witchcraft) alongside Ross Nichols, progressed this idea in the 1950s, and by the mid-1960s it became known as the witch's wheel of the year. In 1974, Aidan Kelly adapted the festival names, drawing on Celtic, Germanic and Anglo-Saxon names for inspiration.

In essence, the wheel of the year arises partly from ancient history and partly from a modern need to reconnect to nature. Just as an almanac marks the seasons, the Sabbaths anchor celebratory dates into those seasons.

Please note that some dates vary depending on the year.

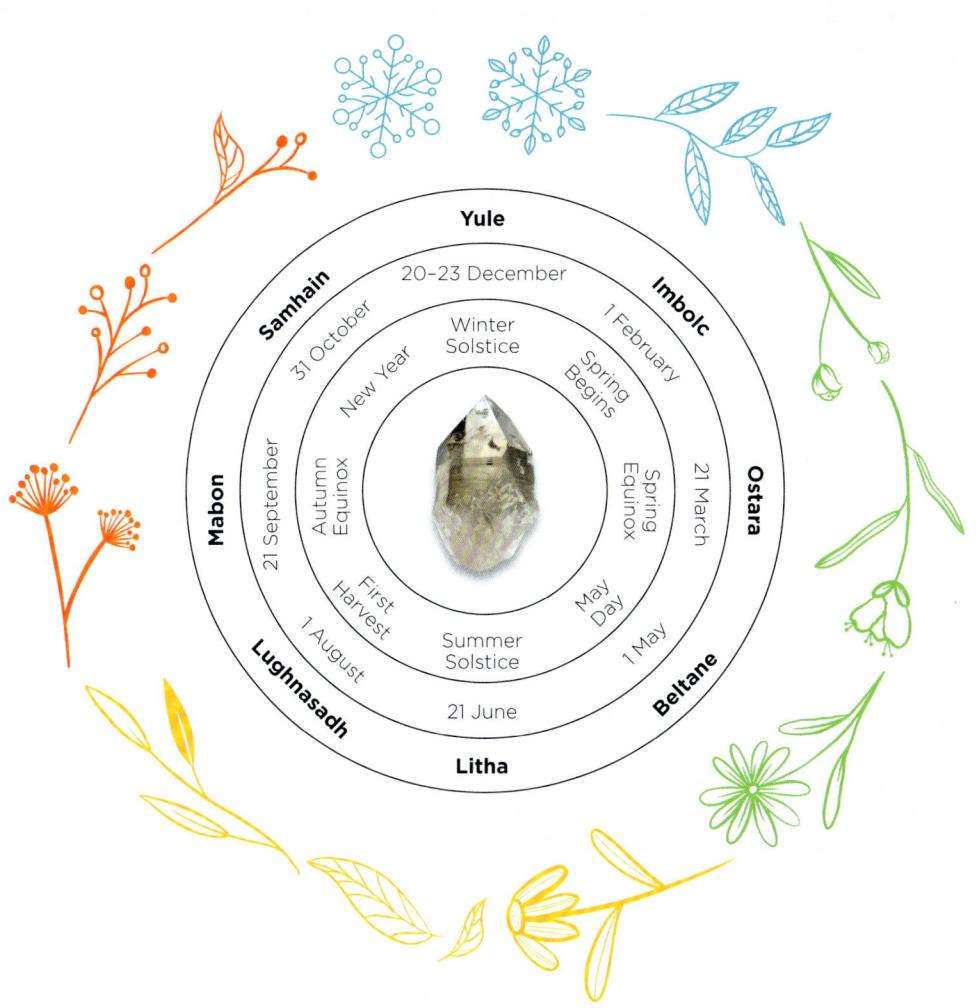

Yule

20–23 December

Samhain

31 October

Imbolc

1 February

Winter
Solstice

New Year

Spring
Begins

Mabon

21 September

Autumn
Equinox

Spring
Equinox

21 March

Ostara

First
Harvest

May
Day

1 August

Summer
Solstice

1 May

Lughnasadh

21 June

Beltane

Litha

Seasonal Celebrations Throughout the Year

MONTH	TYPE OF CELEBRATION		
	GENERAL	BUDDHIST	CHRISTIAN
JAN	New Year's Day World Religion Day Martin Luther King Jr Day Australia Day Lunar New Year*	Mahayana Buddhist New Year Bodhi Day*	Epiphany
FEB	Groundhog Day Valentine's Day Lunar New Year*	Parinirvāna Day* Māgha Pūjā*	Lent*
MAR	International Women's Day	Avalokiteśvara's Birthday* Māgha Pūjā*	Lent* Palm Sunday* Good Friday* Easter Sunday*
APRIL	April Fools' Day Earth Day	Theravada Buddhist New Year* Vaiśākha*	Palm Sunday* Good Friday* Easter Sunday* Ascension*
MAY	Memorial Day* (US)	Vaiśākha*	Ascension* Pentecost*
JUNE	Pride*		Ascension* Pentecost*
JULY	US Independence Day Bastille Day	Asalha Pūjū* Vassa *	
AUG		Ulambana* Vassa*	
SEPT	Labor Day (US)*	Vassa*	
OCT	Black History Month	Pavarāna*	
NOV	Day of the Dead Veterans Day/Armistice Day Thanksgiving	Pavarāna*	All Saints Day First Sunday of Advent*
DEC	New Year's Eve	Bodhi Day*	First Sunday of Advent* Christmas Day

This table highlights some of the best-known dates in the diary. Some of these are connected to the Gregorian or solar calendar, some the lunar calendar, and others use a mix of both. This is why some dates show up once, while others appear several times.

* Date varies depending on the year.
** Dates change annually and will partly depend on the form of Islam that is practised.

TYPE OF CELEBRATION			
HINDU	ISLAMIC	JEWISH	SABBATH
Makar Sankranti*	Ramadan**	Holocaust Memorial Day Tu Bishvat*	
Makar Sankranti*	Ramadan** Eid ul-Fitr**	Tu Bishvat* Purim*	Imbolc/ Candlemas
Holi* Makar Sankranti* Rama Navami*	Ramadan** Eid ul-Fitr**	Purim* Passover*	Ostara*/ Spring Equinox*
Rama Navami*	Ramadan** Eid ul-Fitr** Hajj** Eid ul-Adha**	Passover*	
	Hajj** Eid ul-Adha** Islamic New Year** Ashura**	Shavuot*	Beltane/ May Day
	Hajj** Eid ul-Adha** Islamic New Year** Ashura**	Shavuot*	Litha*/Summer Solstice* Midsummer*
	Islamic New Year** Ashura** Mawlid**	Tisha B'Av*	
Krishna Janmashtami* Ganesh Chaturthi*	Mawlid**	Tisha B'Av*	Lughnasadh/ Lammas
Krishna Janmashtami* Ganesh Chaturthi* Navratri and Durga Puja*	Mawlid**	Simchat Torah* Rosh Hashanah* Yom Kippur* Sukkot*	Mabon*/Autumn Equinox*
Navratri and Durga Puja* Diwali*		Simchat Torah* Rosh Hashanah* Yom Kippur* Sukkot*	Samhain/ Halloween
Diwali*		Hanukkah*	
	Ramadan**	Hanukkah*	Twelve Days of Yule Yule/Winter Solstice*

Working with Crystals Through the Year

People are attracted to crystals for two reasons: their beauty and their stabilizing energy. If you were to put a crystal under a microscope, you would see that its molecules are arranged in an ordered, repetitive way. This consistency gives crystals a very stable frequency, meaning they have a low state of entropy (disorder): they don't change when they are subjected to stress.

In contrast, the human body has different atomic structures in different areas, so we are made up of multiple different frequencies, meaning our state of entropy is much higher, even on a normal day. When stressors enter our life, we become more reactive. In order to be happy, healthy and have positive emotions, we need a high frequency. When our frequency is reduced, our emotions become more and more negative.

Consider two tuning forks that are tuned to the same frequency. When we strike one hard, creating a lot of energy (high amplitude), and then we ring a second gently, the higher amplitude will dominate and influence the lower amplitude tuning fork, raising its frequency. Although the results are more subtle, when we bring crystals into our lives the same happens. There is a crystal for nearly every emotion, thought or experience we might encounter. This book highlights those that support us during the complicated seasonal and astrological energies through the year.

SUPPLEMENTARY CRYSTALS

If there is a month whose energies you do not enjoy, you can bring in extra crystals to amplify your own energies, therefore easing the seasonal effects:

- Wear your birthstone.
- Use a crystal that matches the element of the season you were born into.
- Carry a root chakra (see page 20) crystal you're drawn to.

The Energies of the Seasons

The Crystal Almanac is a journey through the energies of the year, exploring the influence of the elements, deities, astrology and lunar cycles.

ELEMENTS AND DEITIES

In the global north there are four seasons in a year, each three months long. All four have a correlating element that reflects what is going on in nature at that time. As we move through a season, the energies of that element increase, often intensifying its themes, before finally dissipating. This is also the case with the astrological seasons (see page 14).

There are also goddess energies at play. These deities are seen as an archetype of a specific energy. One goddess I refer to throughout is the triple goddess. She is generally viewed as a trinity or having three distinct aspects, each of which represents a stage of the female lifecycle: the Maiden, the Mother and the Crone (the Crone is often called the Wise Woman nowadays, which is the name I will use from now on). Each season is connected to one of those aspects and the underlying energies given off at that time.

SEASONS	ELEMENT	THEMES	TRIPLE GODDESS
WINTER	Air	Introspection, contemplation, stillness, rest	Wise Woman: Wise, intuitive, insightful, lived experiences
SPRING	Earth	Planning, growth, the physical world	Maiden: Joyous, youthful energetic, playful
SUMMER	Fire	Progression, achieving, exploring, maintaining, supporting	Mother: Growing, creating, caring, supporting
AUTUMN	Water	Harvesting, evaluation, emotional healing	Wise Woman: Wise, intuitive, insightful, lived experiences

Each element has correlating locations, animals and herbs. The table below lists these, as well as the crystal and grid shapes that represent each one.

ELEMENT	LOCATIONS	ANIMALS	HERBS	CRYSTAL SHAPES	CRYSTAL GRIDS
AIR	Libraries, aeroplanes and skyscrapers	Bees, birds and dragonflies	Sage, mint and dandelions	Points, towers, clouds and feathers	Fibonacci spirals
EARTH	Caves, forests, country paths, fields, stone circles, ley lines, homes, cities and buildings	Bears, horses, lambs, deer and earthworms	Lemon balm, chamomile and liquorice root	Raw, cubes, carvings of animals, plants and skulls	Flower of Life, hexagon and square
FIRE	Kitchens, deserts and the sites of fire ceremonies	Fireflies, foxes, snakes and big cats	Cayenne pepper, ginger and nettles	Flames, pyramids and incense holders	Cubes or triangles
WATER	Bath, showers, swimming pools, rain forests, rivers, lakes and oceans	Fish, pond and river wildlife; marine mammals	Rose, milky oats and marshmallow root	Water droplets, crystal cups and water mammal carvings	Water element crystals and shells

ASTROLOGY

There are twelve signs in the zodiac, each with different properties and characteristics. The astrological year starts with the Aries season on 21 March. Each season then starts mid-month, taking up just over 30 days. During that time, our experiences and perception of events can be affected by the themes of that sign. This shows to us areas of our lives that require our attention. As we move through the year, you will see the star signs work together to progress our creative cycle, from start to finish. We also alternate between signs that require lots of energy, and signs that ask us to be still and calm. Some signs offer healing or balance, while others give us time to check in with our goals and make sure our plans are built on solid foundations.

THE MOON

It takes the moon just under 28 days to complete one Earth cycle, but just over that to complete all of its lunar phases. Phases refer to the amount of the moon we can see during different aspects of its cycle. There are eight phases: the new moon, waxing crescent, first quarter, waxing gibbous, full moon, waning gibbous, third quarter and waning crescent. Each phase offers unique properties, but the two main phases people tend to focus on are the full moon and the new moon.

NEW MOON

This is the start of the moon's new phase, when only a small amount of it is lit by the sun. It's the perfect time to start something new, make plans and carry out some manifesting.

FULL MOON

At this time of the month, the moon is fully lit by the sun, so you might feel a metaphorical spotlight appear on areas of your life that really need your focus. It is also the time when we can trust that we have created enough momentum for our goals, so we can slow down and turn our attention inwards, focusing more on healing, until the cycle begins again.

The lunar cycles are a big topic, so if you want to follow them more closely, head to my website gemmapetherbridge.com/moon for more information, including a list of lunar dates and their relevant crystals.

Choosing Your Crystals

The Crystal Almanac is arranged by month. Each month, I suggest several crystals that relate to the available energies of that time of the year, as well as possible crystal shapes to look for. I recommend taking the time to review my suggestions and choosing the crystal specimen you would most like to work with.

No two crystals are the same. Their energies depend on where they were found, their age and the other rocks and minerals found around them. Each crystal has its own unique colour tone, detail and shape. When you are browsing in a crystal shop or looking for crystals online, pinpoint a few that you like the look of. Once you've made your selection, complete the following steps:

1. Close your eyes and take some deep breaths to calm your mind.
2. Recall your intention: to select your ideal crystal from those available.
3. Then open your eyes. The crystal you look at first is the right one for you. It's that easy!

Programming Clear Quartz Crystals

When you programme a crystal, you give it an intention: the energy of the thing you want help achieving. If you want to keep your crystal collection small, you can source a Clear Quartz crystal and programme it each month with the properties of the crystal you would like to work with for that month.

Light can travel through Clear Quartz, so it holds all the colours of the rainbow inside, and therefore all themes we might need support with. This makes it easy to programme so it can support us in any way we want.

Simply follow these steps, which take less than 3 minutes:

1. Place your crystal in your non-dominant hand, close your eyes and take a few long, deep breaths.
2. Focus on your breaths and take note of the crystal's texture – this helps you connect to it.
3. Visualize how you would like the crystal to support you. You need to imagine you have already achieved that wish. Let yourself feel the emotions connected to that achievement and allow those emotions to pass through your hand and into the crystal.
4. Take a few more deep breaths and open your eyes.

When your crystal has finished its assigned task, cleanse it (see the next page) and then programme it with your next request.

Cleansing and Charging Your Crystals

Crystals can pick up the energy of the environment they are in, as well as of the people who have interacted with them. It is important to keep their energy pure to allow the properties of the crystals to fully come through. Cleansing is a popular practice that many crystal collectors perform, as it helps you build a closer connection to your crystals.

You don't have to perform this practice too regularly as a crystal's state of entropy is so low (see page 12). I recommend cleansing your crystals four times a year to welcome in each new season. That said, crystals we ask to take on negative energy should be cleansed more frequently in order to remove that energy from that crystal.

Below are three cleansing techniques that can be used on any crystal. These practices only take a few minutes. While you are performing them, make sure you keep the following intention in your mind: *I am removing unwanted energy from this crystal.*

Saging: Light a sage stick. After a moment, blow it out and allow the smoke to pass around all areas of the crystal.

Sound: Sound chimes, bowls or music over the crystal to use one frequency (sound) to clear another frequency (the crystal's vibration).

White light: Cup your crystal between both hands and imagine white light coming from your hand, surrounding the crystal.

CHARGING CRYSTALS

Once your crystals are cleansed, you can display them in direct view of the sun (if they are sun safe) or moon for a few hours. This gives them a chance to take on the energies, heightening their natural properties.

Crystal Shapes

Did you know the shape of a crystal adds to its properties?

Personally, I only purchase a shaped crystal if it complements or enhances the crystal's properties. For this reason, I sometimes suggest specific shapes throughout this book. Nevertheless, if you don't like the shape I've recommended, you don't have to work with it. It is more important that you work with crystals you love being around and feel drawn to. For example, many people only want to work with raw crystals, which is absolutely fine. Below is a table listing all the shapes you will find mentioned in this book and their meanings.

Raw
The natural formation, perfect for display

Clusters
Support communities

Tumbles, nuggets
Polished small crystals, ideal to carry

Chips
Polished crystals, smaller than tumbles. Often used in rituals or for decoration

Slab
A slice of a crystal that creates a natural display base for other items

Towers, points
Direct energy up into the universe

Double termination
Dispels stagnant energy

Harmonizers
Balance energies

Palm stones, worry stones
Calm, comfort and help to deepen meditation

Spheres
Cultivate feminine energy

Cubes
Cultivate masculine energy

Pyramids
Represent duality

Egg shapes
Support new beginnings, birth and motherhood

Hearts
Support relationships and loving kindness towards ourselves and others

Flames
Symbolize aspects of the Fire element

Skulls, alien skulls
When a crystal takes on human form, we build a greater relationship with it

Sacred geometry and Platonic solids sets, Fibonacci spirals and merkabas
Embrace the mathematics of nature, enhancing the energies of the crystals

Shapes, including angels, moons, water droplets and leaves
Symbolize the character the shape represents

Wands
Command and direct energy

Mirrors
Are tools for the psychic skill of scrying

Jewellery
Allows us to wear crystals on the body

Crystals and the Chakras

Some basic knowledge of colour therapy will help you to start to understand the properties of crystals. For example, fiery-coloured crystals increase our vitality, ignite our passion and help us build willpower, while blue crystals calm a bad temper, cool the nervous system and help us communicate better.

A knowledge of the chakras can also further deepen your understanding. Originating from the Vedas (ancient Indian texts that make up some of the earliest Hindu teachings), the chakras are part of our 'energy body'. These discs of energy (the name comes from the Sanskrit for 'wheel') are positioned all around us, and each has a different role. Collectively, they support all aspects of everyday life.

When a chakra is depleted, the energy running through it is reduced, affecting the areas of our lives highlighted in the table opposite. To help correct this, we can use a crystal corresponding to the chakra's colour to heighten its energy. Supporting our energy body then supports our physical body and our general life experience improves.

The seven best-known chakras run up the spine. These will be mentioned throughout the book, so it's important you know where each one is positioned (see image) and how they affect our creativity or goal-setting process and, in turn, the seasons of the year.

Crown Chakra

Third Eye Chakra

Throat Chakra

Heart Chakra

Solar Plexus Chakra

Sacral Chakra

Root Chakra

CHAKRA	CRYSTAL COLOUR	EFFECTS OF DEPLETED CHAKRAS	ADDITIONAL AREAS IT SUPPORTS
CROWN CHAKRA	White and transparent	No direction, or not receiving inspiration.	Meditation, making a connection to something greater, supporting depression and learning difficulties.
THIRD EYE CHAKRA	Purple	Receiving so many ideas that you feel overwhelmed and don't know where to start, or you start too many new projects at the same time.	Intuition, concentration, exam support, relieving poor judgement and lack of focus.
THROAT CHAKRA	Blue	Finding it difficult to progress ideas, first to yourself and then through discussion with others.	Improving public speaking and all forms of communication and creative expression. Learning to speak-up, being misunderstood, being too secretive or not being a good listener.
HEART CHAKRA	Green and pink	Telling people about your ideas but not falling in love with them enough that you feel motivated to pursue them.	Developing love, romance, self-care, compassion and empathy. Supporting forgiveness, trust issues, bitter, hateful and intolerant emotions.
SOLAR PLEXUS CHAKRA	Yellow	Finding it hard to get your projects off the ground because of a lack of confidence, imposter syndrome or lack of vitality.	Supports positivity, self-esteem and gut instincts. Aids feeling powerless.
SACRAL CHAKRA	Orange	Your energy runs out or you lose interest before a project ends.	Supports relationships, emotions and creativity. Aids trauma, low libido, fear of intimacy, and isolation.
ROOT CHAKRA	Red, black, brown, metallic and earthy	Once you have finished a project, you shy away from telling others about it.	Supports financial abundance. Aids fears, anxieties and instability. Grounds us and offers safety and security.

As we evolve, more chakras become available to us. These four additional chakras will be discussed later in the book.

HIGHER CHAKRA	CRYSTAL COLOUR	LOCATION	AREAS THEY SUPPORT
SOUL STAR CHAKRA	Magenta	1 metre (3 feet) above the head	Our connection to our higher self (our soul).
HIGHER HEART CHAKRA	Turquoise	Between the heart and throat chakra	Expanded heart energy.
CENTRAL SUN CHAKRA	Gold and yellow	Over the solar plexus	Who we aspire to be – the wisest aspect of our personality, free from negative egoic traits.
EARTH STAR CHAKRA	Earthy colours with a sparkle to them	1 metre (3 feet) below the feet	Our connection to the Earth and every plant, tree, animal and person we share it with.

THE WHEEL BEGINS

Now that you're equipped with all the background information you need, we are ready to start our progression through the next twelve months.

For each month I'll start with a brief overview of that month's energies so you know what to expect. Then I explore that month's Sabbath celebrations or the season's element so you can see just how closely we live alongside the elements, and how they influence what is going on around us.

We then delve into the astrology and moon phases, which provide us with specific themes to focus on each month, alternating between months where we progress and grow, and months where we relax and look inwards.

To help navigate each month, you will be introduced to one or more deities who support the month's themes. And of course, I will be recommending a beautiful collection of crystals and illuminating how you can use these as seasonal tools to get through the year.

So, as the clock strikes midnight on New Year's Eve and we step into January, our crystal almanac journey begins...

January

MIDWINTER: **STILLNESS, RESTORATION & CONTEMPLATION**

Midwinter is a time of deep introspection and healing. The festivities are over, the nights are long and we are resting after all the celebrations. The animals are hibernating and the plants have receded into the ground. Many of us arrive at the beginning of January feeling exhausted and running on empty.

However, Mother Nature has a special intention for this month, one that the modern world has pushed aside: to permit us to stay inside, recharge and embrace the healing energies that are available to us at this time.

This is a chance for you to do a life audit and allow winter's insightful Wise Woman energies to help you understand where you might be holding yourself back. January gives us the time and space to consider our mental health and heal so that, later in the year, when we want to focus on our aspirations and goals, we are free of past wounds. Unburdened after this reflection, we will allow ourselves to be even more ambitious and strive for even greater success.

This month I show you how crystals can release energy around trapped emotions and old traumas without having to spend time reliving your hardest memories. The crystals and their collective energies will create a space for transformation and self-discovery. I also introduce you to the Buddhist goddess Quan Yin, who helps us connect to the heart chakra to catalyze this healing.

January facts
Colours: Black and dark blue
Crystal shapes: Raw crystals and hearts
Seasonal chakra energies: Root, sacral and heart chakras
Animals: Bears, wolves and owls
Seasonal shrine/altar: New Year decorations, bells, black and blue candles
Intention setting/affirmation: *I am ready to heal*
January birthstone (traditional and modern): Garnet
Birthstone properties: Fertility and physical abundance

JANUARY ELEMENT: **AIR & WINTER**

The idea that life begins and ends with a breath helps us understand just how significant the Air element is.

We are surrounded by this element. It might not be seen, but we know it's there. Its invisibility mirrors the Air element's properties: ideas, inspiration and intuitive insights. It is also symbolic of the clouds and birds in the sky. The night sky view of the universe above us shows us how Air shifts into the ether, which invites us to consider the possibilities of what else might be out there.

During the winter months, while nature enjoys its slumber and the winds move between the empty trees, most people find themselves spending most of their time indoors, enjoying Air element activities such as reading, learning and contemplating the past.

All of these reflective activities show just how much we spend the winter months in the higher two chakras. These are the third eye chakra, which represents our search for knowledge, improved mental health and psychic skills, and the crown chakra, which supports our spiritual development.

Intuitive messages are received into the body through the crown chakra and are interpreted in our third eye, so it's important to keep that line between the two chakras open.

Finally, angels have long been seen as Air element beings. With this energy at its most potent, this is the perfect time to build a connection to them.

CRYSTALS FOR WINTER'S AIR ELEMENT

Knowledge and learning: Rainbow Fluorite

Softening negative mind chatter: Pink Lithium Quartz

Spiritual development: Clear Quartz

Intuitive development: Amethyst

Angelic communication: Celestite

JANUARY ASTROLOGY
& FULL MOON

We start the year with Capricorn season, whose hardworking energies see us through the festivities and help inspire our New Year's resolutions. From 20 January we enter Aquarius season and find ourselves in a fixed Air element sign, and at the height of the Air element season.

Even though January is recognized as the start of the new year by many, Mother Nature and astrology see winter as the end of the cycle. Because Aquarius is the eleventh of the twelve star signs, it is full of potential and new ideas. After all the healing we have done through the winter months, it asks each of us to consider: *Who am I now and what should fill the void?* Witness this month how those answers slowly come via intuitive insights. Then, as nature starts to wake up next month, so will that change within you.

JANUARY'S FULL MOON

This month's full moon is known as the Wolf Moon. Think of the image of the wolf howling, framed by the full moon as it communicates to its pack. The sad reality is that the wolves are known to howl more in January, not for communication but because they are hungry. It's an indication of how animals, along with humans, may wish that January would pass quickly. The image of a wolf also reflects how Aquarius energies make us behave. As the wolf howls into the air, what will fill the space? What will it summon? In our case, we might ask ourselves as we rest and recover, what will fill the space once we are healed? This is what Aquarius wants us to consider.

JANUARY GODDESS: QUAN YIN

NAME AND MEANING
Alternative names for Quan Yin are Kuan Yin, Guanyin
and Kannon. Her name means 'observing the sounds
(or cries) of the human world'. Respected throughout
East Asia and revered in China, Quan Yin is thought of
as a female equivalent of Buddha. She has multiple titles
– she even has a masculine aspect, as well as thirty-three
manifestations. These relate to the countless ways in which this
goddess of mercy might appear to us to offer support and wisdom.

CHARACTERISTICS
Believed to have been a princess who chose a spiritual path over royal
life, Quan Yin reached enlightenment, but as she stepped into heaven she
'heard the cries of the world' and decided to become a Bodhisattva instead,
choosing to support sentient beings until all had reached enlightenment.

Often depicted standing on her dragon, Quan Yin is known for her profound
compassion and ability to inspire forgiveness. Her power comes in the subtle
form of grace, showing us that profound change does not have to be loud and
forceful. Instead, it can be realized through understated, unseen acts such as
opening our hearts to others, detaching from hurt and pain or choosing to
step away from complicated situations so we can find inner peace.

TEACHING
Quan Yin reminds us of the importance of balancing our karma so that the
path to enlightenment is possible. She offers lessons that open the higher heart
chakra that, once open, allows us to communicate with greater sensitivity and
operate in the world with intuitive understanding. To welcome her in, just say
her name; repeating her name multiple times offers healing.

SYMBOLS OF QUAN YIN
A pink lotus flower, a water jar, willow, doves and dragons

JANUARY AIR ELEMENT CRYSTAL:
PINK LITHIUM QUARTZ

CULTIVATING GOOD MENTAL HEALTH

Winter is the perfect time to work with the energies of Air element crystals. Most of them have a transparent appearance. With its translucent nature and its ability to support our mental health, Pink Lithium Quartz is your perfect winter crystal.

If we spend the majority of January in deep contemplation, negative mind chatter can arise. Whether your negative thoughts are connected to past traumas, or your contemplations have led you to understand where limiting beliefs, fears and anxieties might be holding you back, Pink Lithium Quartz helps to soften those thoughts and emotions, releasing the energy around them so that healing is possible.

Transparent crystals also focus on long-term change, so Pink Lithium Quartz wants to cultivate lifelong good mental health. The colour pink also offers heightened levels of compassion, which help progress the healing process.

HOW TO USE IT

If you would like January to become your main healing month, I would recommend treating yourself to a Pink Lithium Quartz necklace. You can wear this annually as a tool to lift your emotions. Personally, I've found that my necklace has also become a talisman; any time I become aware of negative thoughts, I put it on and find myself motivated to keep going.

Additional properties: Develops clairvoyance, softens the impact of bullying, protects against negative words

Source: Brazil

Crystal shapes: Raw, points and jewellery

Best times to work with this crystal: Wednesday, afternoons and evenings

JANUARY ASTROLOGICAL CRYSTAL: MOLDAVITE

CREATIVE THINKING

When you have been doing your healing work all month, and you get to Aquarius season in the second half of the month, it asks you to look ahead and decide what you want to fill the new space in your life with.

Often, when we are presented with the opportunity to create something new, we need to think outside the box and explore ideas we might not have considered before. Uranus is the planet associated with Aquarius. It is one of the outer planets that has a greater connection to other galaxies, so its energies can offer us unique otherworldly insights and ideas. To welcome this kind of work we need to look for crystals with galactic properties.

Moldavite is a 15-million-year-old tektite that was created when an asteroid hit what is now known as the Moldau River area, in the Czech Republic. Technically a natural glass, it was formed by the sudden extreme heating of sand at the impact site. Forged from something travelling through space, it can now offer us alien concepts.

HOW TO USE IT

Hold your Moldavite specimen in your dominant hand for 5 minutes each day while you sit or lie down. After a week, you can increase the time to 10 minutes, adding 5 minutes each week until you can easily sit with it for 20 minutes per day.

After a few days of connecting with Moldavite you might start receiving intuitive information as you hold it. Others will be guided to new information or experiences at unexpected times throughout the week.

Moldavite is a very energetic crystal; even those who don't normally feel crystal energy often feel this one. Therefore, I recommend increasing your time with it slowly.

Additional properties: Exploring abstract concepts, creating original ideas and accelerating spiritual development

Source: Czech Republic

Crystal shapes: Raw and jewellery

Best times to work with this crystal: Monday and Sunday, dawn

JANUARY GODDESS CRYSTAL:
WATERMELON TOURMALINE

FORGIVENESS

January can energetically hold us so we can truly understand and process the lessons behind our biggest traumas. In those moments the emotions around specific situations can be released, allowing us to start moving forward.

When we are in trauma our bodies can take on unhealthy emotions that can leave us in a state of flux, making it hard to move forward. Crystals can lift these energies away from the body, easing their effects and helping us to forgive, or consider the idea of forgiveness. From there we can start looking forward and the effect those events have had on us can start to be healed.

Also called bi-coloured Tourmaline, Watermelon Tourmaline consists of Pink Tourmaline encapsulated inside Green Tourmaline. Together, these are the colours of the heart chakra (see page 21), making it the ideal crystal representation of Quan Yin's compassionate energy. Working with it, you will witness how the grace it exudes dwarfs its small size.

A profound healer for the heart chakra, this crystal sends its pink energy into our hearts, offering us the self-care and compassion we need to overcome past events. The Green Tourmaline then invites deeper insights so that we can find the lessons behind what has happened, overcome and even reach forgiveness, enabling us to fully heal and focus on the future.

HOW TO USE IT

Treat yourself to a Watermelon Tourmaline necklace with a chain that allows your crystal to sit over the heart. Wear it regularly so it can slowly lift any held emotions that are there. On days that feel harder, hold your crystal in your dominant hand and recite Quan Yin's name three times to strengthen the healing.

You will find this crystal's work is subtle, so you might not notice daily changes, but after a few weeks you will look back and see the emotional and physical changes that have been made.

Additional properties: Balancing the heart chakra, supporting grief, cultivating a connection to the animal kingdom

Sources: Africa, Brazil and the USA

Crystal shapes: Raw and jewellery

Best times to work with this crystal: Friday, before bed

JANUARY CRYSTAL OF THE MONTH: BLACK MOONSTONE

SHADOW WORK

The long nights of midwinter allow us to review where we are and where we are going. We can explore and take time to heal even the most complex emotions. This is called shadow work, and it involves the healing of the most difficult human emotions, such as guilt, shame, regret and distrust.

Black Moonstone is the perfect visual representation of this work. Although black in colour, it has a depth that depicts these hidden aspects of ourselves. And there are also flashes of sparkle, which signal to us that working with it to release the energy around these emotions will have a profound outcome.

HOW TO USE IT

I invite you to put your Black Moonstone on display every January so it can remind you to take time to look inside and understand what needs healing. All Moonstones specialize in helping us move through different stages of life, so displaying your crystal will allow it to support you through any transitions this valuable work has for you.

January is also the perfect time to journal. The best time to do this is first thing in the morning. I invite you to write three pages answering the question: *How am I feeling today?* Three pages sounds like a lot, but with practice you will find that after writing down your initial feelings, deeper, more insightful and hidden emotions have a chance to come through.

After journalling, meditate holding a small Black Moonstone specimen in your non-dominant hand. Your intention is to spend no more than 10 minutes intuitively placing your crystal over different areas of the body, to allow your body to release any emotions that come up when journalling.

Once the work is done, make sure you cleanse your crystal so it can release any negative energies it has taken on (see page 18).

Additional properties: Balances the earth star chakra, relieves nightmares, supports witch-wound healing (that is, memories of past life persecution, which can show up in the form of nightmares and fears)

Source: Madagascar

Crystal shapes: Freeforms, spheres, palms and worry stones

Best times to work with this crystal: Monday, evenings

JANUARY ACTIVITIES

1. NEW YEAR'S CRYSTAL RETREAT

Why not welcome in the new year with an at-home day retreat? This gives you the time to carry out all the self-care routines your body will be desiring after December's celebrations. The festive season, followed by the New Year celebrations, accumulates an abundance of optimistic, hopeful energy. This high-frequency energy makes it easier for us to connect to our intuition, so the start of January is a perfect moment to receive guidance for the month ahead.

If you like the idea of using January for healing but you don't know where to start, use this retreat day and its potent intuitive energies to guide you. If you are not naturally intuitive or would like to heighten those skills, bring in crystals to make it easier for you to receive guidance.

INTUITION BODY LAYOUT

1. Lie down and place two Black Tourmaline specimens under each foot. Place a Blue Lace Agate chip over the throat and a double-terminated Amethyst point over the third eye. Finally, place a Clear Quartz point above your head on your pillow, with the point facing towards the head.

2. On the inhale, connect to the Clear Quartz, then the Amethyst, Blue Lace Agate and Black Tourmaline crystals. On the exhale, track back the other way. Moving your awareness up and down like this enhances the line of energy going through the body, so you are more likely to receive intuitive messages.

3. Do this for 12 minutes on the morning of your retreat. Over lunchtime, pen some questions you would like to ask your intuition. In the afternoon do the same activity again, this time thinking of your question on the inhale and intuitively allowing the answer to come on the exhale.

Intuitive guidance can be received through any sense, so the answers might be subtle, and you might have to ask a question several times before receiving a reply. Start with easier questions, before asking more important ones.

2. CRYSTALS, QUAN YIN & DRAGON HEALING

After a month of emotional healing, you might need an energetic refresh. Take an evening to yourself at the end of the month to welcome in Quan Yin and her dragon. Her deep compassion can lift any lingering emotions from the body and her dragon can blow away any unwanted energies from the home.

First, gift yourself a new crystal that for you symbolizes joyful and happy energies – yellow crystals are a good option for this.

INVITING QUAN YIN

1. In a relaxed, meditative state, close your eyes and recite three times:

    I'm proud of the work I have done,
    I'm now ready to release and start anew.
    I welcome in the new me.

2. In your mind's eye, imagine Quan Yin in front of you, placing a hand over your heart. While she does this, see a ring of pink light around your body, in line with the heart. On each exhale the ring gets bigger, until it's larger than your home. When that happens, say to yourself:

    Release, release, release.

3. At that moment, have the intention that any negative emotions are released outside your home and float up into the air. You can now invite Quan Yin's dragon to cleanse your home and refresh its energy. Some people will feel the dragon, others might see it, but you don't have to be aware of its presence to trust that once you have made this request, it is happening.

4. Once you intuitively feel this is done, thank Quan Yin and see her step away. Now take hold of your new crystal specimen and whisper to it all the changes you are hoping this month's healing work will bring to your life. Put it on display as a talisman, a reminder of the work you plan to do now that the healing is complete.

February

WINTER ENDS: **ANTICIPATION, HEALING & DAYDREAMING**

In February, our emotions lift. In the ground, spring flowers begin to emerge. This subtle, hidden shift lets us know that spring will soon be here.

Imagine how excited our ancestors felt as February began. After months of food rationing, they had almost survived another winter. But February is still connected to the Wise Woman, who reminds us that it is too early to let our guard down. It is still cold outside and there is a chance of further frost. Although the end might be in sight, we cannot stop rationing quite yet.

Instead, February is a time to transition from healing our minds to looking at our physical health. When we are not doing that, we are allowed time to daydream. Ask yourself what you would like to have, do and be this year, enjoy the romance of Valentine's Day and see if you can notice the subtle changes that are happening all around you.

This month I will show you how crystals can help us dream and design our ideal year. I will introduce you to the Irish goddess Brigid and we will invite in the healing crystal energies to strengthen our bodies for the year ahead.

February facts
Colours: Light blue and green
Crystal shapes: Flames and wands
Seasonal chakra energies: Root and crown chakras
Animals: Sheep, mice and fish
Seasonal shrine/altar: Brigid Cross, wells, milk and green candles
Intention setting/affirmation: *I release what no longer serves me*
February birthstone (traditional and modern): Amethyst
Birthstone properties: Sobriety, good health, relaxation and the violet flame
(see page 50)

FEBRUARY SABBATH CELEBRATIONS:
IMBOLC & CANDLEMAS

This month we celebrate our first Sabbath. It is generally honoured on 1 February, although some choose to celebrate on 31 January or 2 February, especially those who mark the birth of the Irish goddess Brigid (see page 41).

Imbolc is one of four Celtic cross-quarter celebrations. It marks the midpoint between the winter solstice and spring equinox. Imbolc is a fire festival, so it is a time to light candles and torches to cleanse and invite back the light. Each fire ceremony honours that season's element, and Imbolc connects us to winter's air, so there is symbolic significance to the smoke the fire creates. This makes Imbolc the perfect time to sage your crystal collection (see page 18).

This is also a moment to honour the sense of possibility in the air. We feel relief as winter releases its grip and we begin to look forward to the months ahead. Many who follow the seasons see the start of February as the beginning of spring. On this date, our ancestors would stop their rationing for a moment to rejoice and lift their spirits after months of dark nights, enjoying some light-hearted fun before the hard work of the lighter days began.

Some believe the word Imbolc means 'in the belly', while others offer the Celtic term *oí-melg* ('ewe milk'), indicating 'the time that sheep's milk comes'. Historically, this would have been cause for celebration. By now our ancestors' rations were running low, so the arrival of ewes' milk and its produce would have provided an immediate source of fresh food, which meant they could open their final preserves, dried fruit and herbs to create a feast of fresh lamb.

The Christian festival Candlemas was also celebrated at this time. Candles were lit to mark the purification of the Virgin Mary. These candles reminded people to clear out the old and prepare for a new start.

CRYSTALS TO DISPLAY FOR IMBOLC

Messages of hope: Peridot

Fertility: Peach Moonstone, Garnet and Ruby

FEBRUARY ASTROLOGY
& FULL MOON

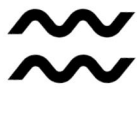

We start this month still in deep contemplation with Aquarius. Then, as we move through the month and nature wakes from its hibernation, we enter the season of Pisces on 20 February. This is a mutable sign, meaning it is flexible and makes changes in our life. This is a powerful message, as Pisces marks the end of the astrological year and the transition into the new one next month.

When this happens, we relax into a time of daydreaming, exploring who we are and who we want to be once spring and the new astrological year begin. This is inspired by Neptune, the planet of Pisces, which is connected to aspiration and dreams.

Pisces is also a Water element sign, so it uses emotions to help us understand how we feel about a situation. When ideas come to the surface this month, it allows us time to contemplate all options as our minds operate like the Pisces fish, swimming in all directions before choosing the preferred direction.

FEBRUARY'S FULL MOON

The most common name for February's full moon is the Snow Moon, but it is also referred to as the Storm Moon and the Hunger Moon. All these names remind us that although winter is drawing to a close, it is not over just yet.

FEBRUARY GODDESS: BRIGID

NAME AND MEANING

The pre-Christian Irish goddess Brigid (Brigit, Brig or Saint Brigid) is the goddess of the sun and returning light. Her name means 'high one' or 'exalted one'. Imbolc marks the date on which Brigid was born. This Celtic goddess is beloved by Irish people and later became a Christian saint, though it is unknown whether she was a real person.

TEACHING

Brigid is a triple goddess, meaning that aspects of her represent all three stages of the divine feminine: the Maiden, the Mother and the Wise Woman. In February, the Wise Woman energy starts to fade away and the Maiden energies of new beginnings ease in before they fully materialize next month.

CHARACTERISTICS

With her connection to the triple goddess, some people believe Brigid is actually three different goddesses all in one, with different spellings of the same name, because there are many different aspects attributed to her. Brigid is a keeper of the hearth and home, the goddess of healing, midwifery, poetry, the smith and the forge.

On Brigid's Eve, 31 January, it is believed that Brigid would move over the land. The head of the house would put out the fire in the hearth and create a carpet of ash in the fireplace. In the morning, the family would check for signs in the ash to see if Brigid had passed through.

Since pre-Christian times, a sacred fire has burned in Kildare, Ireland, to honour Brigid. Priestesses once gathered on the hill there to tend to the fires, invoking Brigid and asking her to protect their herds and to provide a fruitful harvest. Today it is believed that the fire is still tended by Brigid herself, alongside 19 of her nuns.

SYMBOLS OF BRIGID

The Brigid cross, the sacred flame and holy well

FEBRUARY SABBATH CRYSTAL: ELESTIAL QUARTZ

MULTIDIMENSIONAL, MULTIPLE OPPORTUNITIES

February has a sense of possibility: whether it's from the chance of finding love on Valentine's Day, the sense of anticipation that keeps building from the Imbolc celebrations on 1 February, or the daydreamy energy of Pisces, February's possibilities feel endless.

We are in a time when humanity is raising its vibration from a third dimensional reality, represented by the everyday physical world we live in, to a fifth dimensional reality that will subtly allow us to shift our behaviour, so that we become collectively more compassionate towards each other and the planet.

All of this is encapsulated in the energy of Elestial Quartz, also called Skeletal Quartz or Jacare Quartz. This crystal allows us to access other dimensions, including the realms of the angels, deities, our ancestors and other light beings. This is a useful skill, as it allows us to gain insight into new opportunities we might not be aware of yet.

HOW TO USE IT

The idea of shifting our reality may sound like something from a sci-fi movie, but you can view it merely as a simple change in perspective, along with a slight change in your vibration. The energies surrounding us in February will help you see 'outside the box', making change possible. To support this, Elestial Quartz can help raise your vibration. Simply hold a small Elestial Quartz specimen in each hand as you daydream to enable those subtle energy shifts.

Additional properties: Facilitates a connection with our angels and spirit guides, corrects bad decisions. It is an instant manifester

Source: Brazil

Crystal shape: Raw

Best times to work with this crystal: Wednesday or Thursday, before bed

FEBRUARY ASTROLOGICAL CRYSTAL:
RAINBOW MOONSTONE

DAYDREAMING

This month's daydreams will be asking you to imagine who you would be if you lived in a perfect world. For this you need a crystal that has all the colours of the rainbow so you can access all potential ideas and inspirations. It also needs to have the feminine energies of winter's intuitive inspirations and a strong connection to the star sign Pisces via its Water element properties. A flashy Rainbow Moonstone specimen, which is also sometimes called White Labradorite, contains all three.

All dense crystals, including Moonstone, are seen as energetic tools we can use in the physical world to support everyday life. White crystals support our spiritual progression, so having your Rainbow Moonstone to hand will streamline your daydreams this month, helping you pinpoint the most significant ones.

HOW TO USE IT

When daydreaming, try to focus your attention on the rainbow flashes dancing over the surface of your crystal. This simple action calms the mind and helps you connect to the crystal's energies so you can bring forward new ideas. This is a form of scrying, a psychic practice in which we focus on one specific thing to calm the mind so that we can receive guidance, either as intuitive symbols in the item itself, or through intuitive messages.

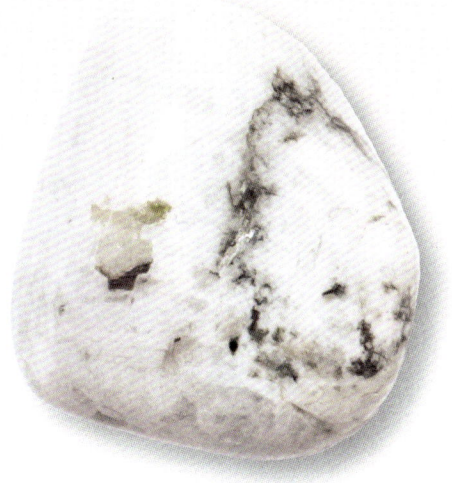

Additional properties: Deepens your connection to the moon, supports change and offers insights

Sources: India and Sri Lanka

Crystal shapes: Freeforms, spheres and moon shapes

Best times to work with this crystal: Monday, dawn

FEBRUARY GODDESS CRYSTAL: AMETHYST

PHYSICAL HEALING

One of the many ways we can work with Brigid is to focus on our physical health. Legend tells of Brigid carrying out miracles for those desperate for support. As the winter months come to an end, the start of February is a perfect time to welcome in any healing energies Brigid would like to offer before we step into the busier months.

For this we can turn to February's birthstone, Amethyst. Perhaps the most famous of healing crystals, its use as a healing stone dates all the way back to ancient Greece. We can internalize this crystal's energy by making a crystal elixir – water that has been infused with the vibration of a crystal. I love working with elixirs because they are portable and cost effective. If you can only afford small tumbles, using them to create elixirs maximizes their energies and will give the same results as working with much larger crystal specimens.

HOW TO USE IT

To make your Amethyst crystal elixir, you need a polished Amethyst crystal, a drinking glass and some bottled drinking water. First, cleanse your crystal (see page 18) and place it next to the drinking glass. Pour the drinking water into the glass. Leave them next to each other for a minimum of 20 minutes, then drink the water.

INDIRECT AND DIRECT METHOD ELIXIRS

The elixir method used above is called the 'indirect method' because it involves placing the crystal next to a glass of water, not in it. Placing a crystal into the water to create an elixir is called the direct method. This technique can only be used with certain crystals, as others can leave residue in the water that can be harmful if consumed. Some crystals will also rust or break down if placed in water. Polished Amethyst can be placed in the water, but to avoid all the above issues, I recommend you only create elixirs using the indirect method. As everything is energy, the crystal's vibration will still reach the water and programme it without any of the risks involved with the direct method.

Additional properties: Relieves old emotional wounds, opens the third eye, plots a course towards a healthy lifestyle

Source: Worldwide

Crystal shapes: Tumbles, chips, flames, hearts, points and wands

Best times to work with this crystal: Friday, mornings

FEBRUARY CRYSTAL OF THE MONTH: ORANGE CALCITE

CREATIVITY

Orange Calcite exudes creative energies. As nature starts to grow back all around us, our feelings of creativity become reinvigorated. We begin to form ideas in our minds that will soon become summer projects.

As we start doing our life's passion, everything begins to come together easily. People sometimes describe this as being 'in our flow'. This is the energy of Orange Calcite. Interestingly, moments of 'flow' are said to replicate the feeling of Buddhist enlightenment, which goes some way towards understanding how important this energy is for us.

If you start working with this crystal now, it will help energize you so you can live in this positive, inspired space all year.

HOW TO USE IT

This is a budget-friendly crystal, so it's worth investing in five small specimens. Store one in your bathroom, moving it closer to your bath water when you want a creative bath elixir (see page 79). Place one over your diary to welcome in lots of creative projects. Then, every week you can perform the following activity with all five crystals.

1. Lie down and place one Orange Calcite crystal above your head and one beneath your feet.

2. Position the third crystal over the solar plexus (see page 20) and the final two in either hand, palms up.

3. Lie with all crystals in place for 10 minutes, focusing on the following intention: *I welcome more creative energy into my life.*

4. When doing this, you might like to place some Clear Quartz tumbled stones around the solar plexus crystal to heighten its energies.

Additional properties: Manifestation, positivity and creative thinking

Source: Mexico

Crystal shapes: Raw, palms, points and double-terminated points

Best times to work with this crystal: Tuesday, early afternoons

FEBRUARY ACTIVITIES

1. **HEALING RETREAT**

The final winter month is potent with the healing energies of the season. With the goddess Brigid offering her healing skills, I now invite you to embrace the end of the astrological year by dedicating a day to deep physical healing so you can maximize February's tone of finality and release any unwanted energies.

The violet flame is a healing energy that we can use to transmute a situation from something negative into something more positive. Its frequency resembles that of Amethyst, making this an ideal vibration to tap into when working with this crystal. We can use it to shift old stuck energies from the physical body and replenish our cells. This is known as spiritual alchemy and shows how we can consciously direct energies to improve our lives.

VIOLET FLAME HEALING
You will need Amethyst chips, an Amethyst point or wand and a Black Tourmaline specimen.

1. Lay out the Amethyst chips in a circle large enough for you to lie in the centre. Lie in the centre of the circle and place your Amethyst wand over your heart as you relax.
2. Imagine an energetic violet flame in your heart. See it giving off a violet smoke that increases with every exhale until it surrounds your body.
3. See the smoke increase until it is a dome of light hiding you from the world. Stay there for 20 minutes. On your exhales, have the intention that you are releasing old energies. On your inhales you are taking in the new positive energy created by the flame.
4. After 20 minutes, visualize the energy coming back to your heart. Take the wand away from your heart and place a Black Tourmaline crystal under your feet for a few more minutes to ground your energies.

2. ELESTIAL QUARTZ VISUALIZATION

Allowing our minds to dream creates new neural pathways in the brain, which opens us up to greater creative thinking, helping us discover new ideas, concepts and inventions. With Pisces' daydreams to guide you, I invite you to go deeper and create your own world. This is a place you can visit any time you want to explore and be guided through complicated, abstract situations.

CREATE YOUR VISUALIZATION

Lie down, holding two Elestial Quartz crystals in each hand. This will help enhance your visualization skills. Ideally you would also place a small Clear Quartz point on the pillow above your head, facing towards your head. Then place an Amethyst over the third eye and a small Blue Lace Agate chip on your throat. These will help you channel the images better.

Now follow the following steps as closely as possible, reading each one before you start so that you understand the full process:

1. Take slow, deep breaths, until you feel relaxed. Imagine walking along a path in a forest. What sounds and smells would you experience? What would the heat from the sunlight breaking through the trees feel like against your skin?

2. You find an ancient tree. You move closer and find a doorway. Open the door and step inside. A spiral staircase appears. Go down, with the intention to become more relaxed as you go. At the bottom you find another door, leading outside. As you step out, you enter your own unique world. Try not to pre-empt what you will see. Instead, allow your mind to create something unique and unexpected, a land that might change each time you visit.

3. Allow yourself to wander around. You might have the intention that you meet someone who can offer insights and guidance. When you have finished, find the tree again, walk back up the spiral staircase and, once you have left the tree, open your eyes.

March

SPRING BEGINS: **RENEWAL, BALANCE & POTENTIAL**

Meteorological spring begins on 1 March. The Maiden goddess has awoken and nature is reviving. The first leaves are on the trees, buds are sprouting and bulbs are starting to bloom. Young animals are in the fields and birds are nesting in the trees, enhancing the feeling of new beginnings.

This month we celebrate the spring equinox, the halfway point between the winter and summer solstice – and a time of balance. After that, the days get longer, naturally lifting everyone's mood and motivating us to go out and spend more time in our community.

With the start of the new astrological year in late March, the energy of new beginnings is everywhere. When March begins we are still with Pisces in its daydreaming. Then Aries offers us the clarity we need to refine those dreams and research our options so we can understand what we want to do.

This month I will show you how the new astrological year asks us to bring more balance into our lives. Together we will honour the new seasons by connecting with ancient Greek goddesses Demeter and Persephone, and I will show you how you can use crystals to harness this energy of fresh beginnings.

March facts
Colours: Pastel pinks, purples, yellows and greens
Crystal shapes: Eggs, pyramids and points
Seasonal chakra energies: Sacral and third eye chakras
Animals: Rabbits, hares and chicks
Seasonal shrine/altar: Bulbs, eggs and hot cross buns
Intention setting/affirmation: *I embrace new beginnings*
March birthstone (traditional): Bloodstone
Birthstone properties: Blood health and our bloodline
March birthstone (modern): Aquamarine
Birthstone properties: Safe travel, connects to the water elements, Atlantis, Poseidon, whales and dolphins

MARCH SABBATH CELEBRATION:
OSTARA/SPRING EQUINOX

Ostara is the vernal, or spring, equinox. It typically occurs on 20 March, but can vary by a day on either side. This celebration marks a moment in time, rather than a date in the diary. Around the world we all acknowledge the exact moment when the sun is positioned directly above the equator, creating a day and night of almost equal length (the word equinox is derived from the Latin for *aequus*, 'equal', and *nox*, 'night'). Then, in a matter of moments, the sun moves again and the northern hemisphere officially moves into spring, while the global south steps into autumn.

This day also marks the start of astrological year. Ancient celebrations to welcome in springtime are present in many cultures, but Ostara was only officially added to the wheel of the year and given its name in 1974. Ostara (or Eostre) is the name of the German goddess of spring and dawn. The word Easter has the same origins.

Eostre contains the root word *eos*, which translates as 'beginnings'. Like spring celebrations witnessed in other cultures, Ostara symbolizes fertility, rebirth and renewal. It offers a day of fun before the agricultural season begins and farmers need to return to the fields to start planting this year's crops.

There are numerous stories surrounding this tradition; some honour the rebirth of the Maiden goddess, others explain the creation of the Easter bunny. Feasting is an important aspect of every celebration. This time of year, the hot cross bun is a favourite treat, and while many see it as a symbol of Easter, its heritage dates back much further. They were actually left as offerings to the goddess Eostre, with the cross representing the crosses of the equinoxes on the wheel of the year.

CRYSTALS TO DISPLAY FOR OSTARA

New beginnings: Selenite and Peach Moonstone

Fertility: Peach Moonstone, Garnet and Ruby

Connection to the natural world: Moss Agate and Garden Quartz

---

MARCH ASTROLOGY
& FULL MOON

We begin this month in the Pisces season, the last sign in the zodiac. I invite you to see if you can sense the subtle shift in energy as we transition on 21 March into Aries, the first sign of the astrological year.

Like its symbol, the ram, Aries can come charging into our lives using its Fire element energy to get us to focus and start working. We enter this sign at the same time as we celebrate the spring equinox, a time when the light wins over the dark and the days become longer. The increase in the sun's energy stokes the fires of Aries, asking you to focus on yourself and your needs.

Fire element signs bring out our passions so we can see our goals more clearly. This month Aries asks us to honour our own needs so we can become our happiest self for those around us. If we are not doing this, Aries shows us how it can lead to frustration. Hone in on any feelings of restlessness, as this shows you where change is needed.

If March is feeling like a butting ram constantly nudging you towards change, choose to keep some Water element crystals with you (see page 165) to calm any emotions or stressors that arise. Alternatively, you can bring in Fire element crystals (see page 109) if you want clearer guidance on where to focus your time.

MARCH'S FULL MOON

This full moon is known as the Worm Moon because it marks the time when worms normally appear above the surface again, signalling to farmers that the soil has defrosted and they can start working the land.

MARCH GODDESSES:
DEMETER & PERSEPHONE

This month we meet two goddesses: Demeter, goddess of harvest, crops, agriculture and fertility; and her daughter, Persephone, the fertility goddess.

NAMES AND MEANINGS

DEMETER	PERSEPHONE
The goddess of harvest and agriculture	The goddess of spring, the dead, the underworld, grain and nature
Also known as Deo	Also known as queen of the underworld, Kore or Cora
Meaning of the name: Earth Mother	Meaning of the name: Female thresher of grain

CHARACTERISTICS

One day, while she was out collecting flowers, Persephone was abducted by Hades, lord of the underworld and brother to Demeter and Zeus. Hades took Persephone into the underworld to be his wife. Having searched for Persephone to no avail, Demeter left Olympus and went to live among the mortals. When she did, the crops started to fail. Afraid for mankind, Zeus told Hades he must free Persephone. Hades agreed, but first he tricked her, giving her six pomegranate seeds to eat. This sealed her fate: because she had tasted the food of the underworld, Persephone would now have to spend a third of each year there as its queen. For the rest of the year, she could return to her mother.

TEACHING

This story explains the creation of the seasons. When Persephone descends into the underworld, Demeter's sadness means everything dies and winter comes. When she returns, Demeter rejoices and nature flourishes.

SYMBOLS OF DEMETER & PERSEPHONE

Cornucopia, wheat and bread; Pomegranate, torch and flowers

MARCH SABBATH CRYSTAL:
BLUE LACE AGATE

BALANCE

The seasons and Sabbaths all have energies and wisdoms we can tap into, but our busy modern lifestyles mean we might miss those messages. We can make the most of powerful astrological days like Ostara, when those energies are at their most potent, so it becomes easier to access their guidance.

On the spring equinox, be aware of your thoughts and feelings. Today is a time when night and day are equal. For a moment, there is a perfect balance. Ask yourself the following questions: *Where in my life am I out of balance* and *What do I need to do to correct this?* Identifying this now will mean you can enjoy the summer months without experiencing burn-out.

All Agate crystals bring us into equilibrium. Working with this crystal can illuminate the areas that need our focus and bring us the wisdom to understand how anything out of alignment can be corrected.

HOW TO USE IT

I suggest you work with Blue Lace Agate because it helps calm the nervous system. Its blue tones also activate the throat chakra, improving our communication skills (see page 21). Introducing more calm and better communication into our lives will help us receive guidance from the equinox.

To harness the power of Blue Lace Agate on the equinox, hold the crystal in your dominant hand and focus on slowly increasing the length of your inhale and exhale. When you feel relaxed, invite your mind to show you images of where you are out of balance. It might show you one or a few areas to focus on. If a few are revealed, ask it to show you which ones to prioritize. Then, ask for some visions indicating how you can correct that imbalance.

Additional properties: Calms anxiety, works through layers of an issue, cools a fiery temper

Source: South Africa

Crystal shapes: Raw, palms and hearts

Best times to work with this crystal: Monday, mornings

MARCH ASTROLOGICAL CRYSTAL:
RED IRIDESCENT AMMONITE

SPIRALLING ENERGY

Unlike the fossils you might see in a museum, Red Iridescent Ammonite (also called Fire Ammonite or Opalized Ammonite) shimmers red with an iridescent sparkle. This is because the fossil's shell has been compressed with enough force for the calcium carbonate to recrystallize and form Aragonite. The Fibonacci spiral in an ammonite represents the sacred geometry of the universe. This shape attracts or manifests things towards us.

The fiery, spiralling energy of this ammonite reminds me of the Aries season. After the quiet winter months, the Aries ram comes bounding in to wake us up and asks us to start taking action. It reminds us that it's time for new beginnings. We have done the healing, made space for new things to enter our life and now it's time to focus on the future.

Aries acts intuitively, often progressing forward without the need for reason. That means change can come in fast this month, and a Fire element crystal like Red Iridescent Ammonite can light the way ahead. This invigorates people who normally like a slower pace of life, so they can match the energies March has to offer. At the same time, its rainbow flash holds the frequency of all the chakras and therefore supports all areas of life, so it can support anything we want to manifest.

HOW TO USE IT

Aries is all about welcoming new experiences. However, to successfully attract new things, we need the flow of energy through our chakras to be open and clear. Over time, our chakras can become blocked, or their rotating discs of energy can slow down or stop. If you struggle to manifest things into your life this might be hindering you.

Red Iridescent Ammonites are the perfect tool to restart the rotation of a chakra. Simply lie down and place your ammonite over each chakra for 3 minutes, starting at the root chakra up to the crown chakra, then moving down the body again to finish at the root chakra.

Additional properties: Abundance, ancient wisdom and past life information

Sources: Canada and Madagascar

Crystal shapes: Raw and jewellery

Best times to work with this crystal: Tuesday and Thursday, early afternoons

MARCH GODDESS CRYSTAL:
GODDESS STONE

DEEPENING YOUR CONNECTION WITH THE GODDESS

Since we explored the story of two goddesses, Demeter and Persephone, this month, it makes sense to introduce you to Goddess Stone (also called Menalite or Fairy Stone), a crystal that has a profound connection to all deities.

A calcium carbonate crystal, Goddess Stone has a smooth, chalk-like surface that resembles both the ancient marble carvings of the female form and modernist sculptures like those by Henry Moore or Barbara Hepworth.

Goddess Stone can help you cultivate a deeper relationship with a deity, first allowing you to identify the deity with whom you would like to connect and then helping you build that connection.

HOW TO USE IT

Find a prominent place in the home to display your crystal, somewhere that acknowledges the important role this crystal will have in your life.

At the start this stone will help to attract the insights you need in order to find your perfect goddess. The guidance you need will begin to present itself to you in the outside world. Once you have that information, sit with your crystal in meditation and it will help deepen your practice so that a connection can be made.

Many people seek a relationship with more than one deity, although it's a good idea to start with one at a time. You can display the Goddess Stone every time you want to develop a new relationship.

Additional properties: Supports the divine feminine, all stages of pregnancy and balances hormones

Sources: Canada, Greenland, the UK and Morocco

Crystal shape: Raw

Best times to work with this crystal: Monday or Friday, mornings

MARCH CRYSTAL OF THE MONTH:
CHERRY BLOSSOM AGATE

PERSONAL GROWTH

March's crystal must honour the beauty of the natural world as well as all the wonderful potential that is available to us at this time of the year. For me, this crystal is Cherry Blossom Agate, also called Flower Agate.

I choose to refer to it as Cherry Blossom Agate because it helps me remember its properties. This crystal guides us so we can 'blossom' into the best variant of ourselves. Unlike other crystals, which can energetically push us onto the right path, Cherry Blossom Agate does it with the subtlety of all Agate crystals. Its Zen-like nature means it can also help you tap into the potential of the spring equinox, allowing you to experience a moment of perfect balance.

HOW TO USE IT

A particularly attractive crystal, Cherry Blossom Agate makes the perfect statement piece when displayed in a place where everyone can enjoy its beauty and benefit from its properties. I recommend sourcing a specimen that also has Clear Quartz druzy forming in it, so the Quartz can amplify its properties. March is also associated with *hanami*, the Japanese celebration of cherry blossoms. This is the perfect time to move your crystal to a prime location in the home, honouring the cherry blossom in your own way and welcoming in new beginnings.

Additional properties: Life purpose progression, confidence, trusting in yourself

Source: Madagascar

Crystal shapes: Towers and flames

Best times to work with this crystal: Thursday, afternoons

MARCH ACTIVITIES

1. GARDENING WITH CRYSTALS

If you love your garden, I'm sure you want to get out there as soon as spring begins so you can start cultivating something beautiful for the months to come. Plants and animals enjoy the energy of crystals as much as we do. Adding a few chips to their flower beds will help create an environment full of the energies needed for the plants to thrive.

This can also be seen as an offering to Mother Nature. Offerings are an ancient practice that require us to first give a gift to the goddess (or the universe) before receiving something in return, such as a plentiful crop. In this case, you might connect to the goddess Demeter (see page 55) or Gaia (see page 125) in your mind's eye as you sprinkle the chips. Acknowledge that the offering is for her, and ask if she can help look after your garden.

PLANT-CRYSTAL AFFINITIES

Over the years my crystal community and I have tested using crystals to support our plants and identified a few crystals that plants seem to enjoy the most. Perhaps it is because the water in the soil is being programmed by the crystal chips, creating a crystal elixir for the plants (see page 46 for more on elixirs.) Alternatively, the energies may be creating an enhanced environment for growth. It could even be a mix of both.

Rose Quartz: Creates a loving, caring environment for plants to grow in.

Amethyst: Creates a calming, healing environment for your plants to grow.

Moss Agate and Cherry Blossom Agate: Both are crystal companions (supports) for the plant kingdom.

HOUSEPLANTS

You don't need a garden to do this activity: you can also add crystal chips to the pots of your indoor plants!

2. CRYSTAL GRIDS FOR NEW BEGINNINGS

The potential in the air at the start of spring makes it the perfect time to make a crystal grid by placing crystals into a shape that, once activated, will carry out a specific task or manifest things in your life. Both the crystals you use and the shape of the grid will manifest the intention of the grid.

Crystal grids draw on the law of intention, which is believed to reward any focused action with a positive outcome, and the law of attraction, which holds that wherever we focus our attention, more of that will come into our lives. They also deploy sacred geometry in the shape of the grid – a Fibonacci spiral in this case – and in the atomic structure of the crystals. This deliberate use of universal mathematics helps create a faster, stronger connection to the things you want to manifest.

HOW TO MAKE YOUR CRYSTAL GRID

1. You'll need a large crystal – I recommend a Clear Quartz tower that has been programmed with your goal (see page 17) – and a collection of crystal chips. I suggest Citrine and Rutilated Quartz for their manifesting properties.

2. Start by deciding where you will display your grid. A ritual space is perfect; it should be a place where the grid is unlikely to be disturbed. Just before you start, cleanse the crystals (see page 18).

3. While thinking about the outcome you want, build the grid. First, put the large crystal where the centre of the spiral will be. All the energy will be channelled to this point. Next, add crystal chips to make a spiral, moving outwards. To activate the grid, trace the spiral from the outside going inwards, using your fingertips. Do this slowly three times while thinking of the desired outcome, then take a breath and do it again three more times, faster.

4. Keep your crystal grid until the intention has been met. If you see that your wish is getting closer, you can create a new grid that reflects your changing circumstance. Remember to mindfully thank the grid before taking it down, and to cleanse the crystals afterwards.

April

MID-SPRING: **PREPARATION, WISDOM & PROSPERITY**

It's the middle of spring. More leaves appear on the trees, flowers bring colour back to nature and bird song is in the air. The month begins with April Fool's Day and a chance to laugh, joke and raise people's spirits. As we spend more time outdoors with friends and family, fun and laughter fills the air.

There are multiple religious holidays this month, but no Sabbath celebration. Instead, farmers are busy working the land and raising new livestock. As life grows, the sun energizes those working in the fields, helping to support their hard work and keep them focused.

In astrology, we move from the fiery passion of Aries to the sensible, pragmatic and earthy energies of Taurus. The spring season is connected to the Earth element, so the two together amplify Earth element concerns – in particular, the need to check that any plans we have are grounded and well thought out.

This month focuses on practical real-life situations, so it asks us to step back and take an objective view of our goals. The universe is offering its wisdom and asking you to consider what needs refining or changing to guarantee success. Together we will seek the wisdom of ancient Greek and Christian deity Sophia, connect to the stones beneath us in our communities and take time to consider our financial security and self-care.

April facts
Colours: Greens and browns
Crystal shapes: Spheres and squares
Seasonal chakra energy: Root chakra
Animals: Elephants, tortoises and caterpillars
Seasonal shrine/altar: Money, leaves, potted plants and cherished possessions
Intention setting/affirmation: *My ideas are founded in wisdom*
April birthstone (traditional and modern): Diamond
Birthstone properties: Clarity, strength, transformation and prosperity

APRIL ELEMENT:
EARTH & SPRING

This season the focus is on the Earth element. From our physical bodies to the Earth's core, our favourite destinations, homes, belongings, the plants and animals we share Mother Earth with, everything is part of the Earth element.

Connecting to this element normally comes more naturally to us than the other elements. It's the unexpected result of long walks, gardening and country pursuits. We are connected to nature when we feel sand under our feet, visit graveyards or natural tourist attractions.

I'm sure you are starting to see how the cycles of the year are in beautiful symmetry with each other. Therefore, it is no surprise that we celebrate the Earth while growth is happening all around. Although we often associate Earth with dense, still structures that are stable and unmoving, this slow growth means that when we look closer there is still movement.

April is also a good time to connect to the Earth element properties of the root chakra and its associated crystals. This chakra represents our fundamental needs: safety, security, food and a roof over our head. It also focuses on financial prosperity.

CRYSTALS FOR SPRING'S EARTH ELEMENT

Connecting to the Earth: Hematite

Connecting to nature: Moss Agate

Root chakra crystal: Garnet

Financial prosperity: Pyrite

APRIL ASTROLOGY
& FULL MOON

The fast pace, progressive energy of Aries moves us into April. The ram is still asking us to rush ahead, putting plans in place to do, do, do. On 21 April Taurus steps in, reminding us that we cannot go at high speed all the time. Instead, it grounds us and suggests we take a step back to reflect and consider which decisions made during Aries season were the most sensible ones, and let go of the rest.

In all areas of life, Taurus is asking us if we are laying the best foundations for success. In essence, its message is symbolic of the oak tree, whose roots grow out as wide as its branches, creating a stable foundation that lasts the test of time. If you feel unsettled in some areas of your life this month, that is Taurus bringing to your attention to where more stability is needed.

Taurus is connected to Venus, the planet of love, beauty and the enjoyment of the material world. That means there will also be a focus on possessions and home life. You might find yourself motivated to create a nicer space to work on your goals, or to landscape your garden in time for summer.

With its connection to the material world, this is also the star sign that focuses on financial abundance, so now is the perfect time to consider your relationship with money and ask yourself if you feel financially secure? If you don't, allow Taurus to intuitively guide you this month so you can work on any financial blocks and start to attract more prosperity into your life.

APRIL'S FULL MOON

With spring flowers appearing everywhere, it is no surprise this moon is called the Pink Moon. Often this is the first full moon after the spring equinox, which is called the Paschal full moon. The date of this moon determines when Easter will be.

APRIL GODDESS:
SOPHIA

NAME AND MEANING

Sophia is honoured in multiple religions, from
Christianity to Wicca to neo-paganism. Known as the
goddess of wisdom, her knowledge is unsurpassed. Many
say she is the wisdom of God, or God's female soul. This
makes her an aspect of the divine, potentially universal
consciousness itself, rather than a distinct goddess.

CHARACTERISTICS

Sophia is also known as a creator goddess who birthed both the female
and male energies into the universe, which could be why she is said to love
humanity so much, and wants to offer her knowledge to relieve our suffering.
Doves are symbolic of this goddess, and represent her overall aim of peace
for humanity. Many also believe Sophia was the dove that told Mary she
would give birth to Jesus.

TEACHING

I'm introducing you to Sophia this month because she also wants to protect
us. She does this by helping us seek the truth. This means she can work
with us intuitively, offering guidance in areas of our lives that are not built
on firm foundations and therefore need to change. You might think that
messages like this would create fear and anxiety, but Sophia's wisdom is
offered in a way that can be received with ease, allowing the recipient to
carry on with hope and optimism for the future.

SYMBOL OF SOPHIA

The dove

APRIL EARTH ELEMENT CRYSTALS:
THE CRYSTALS & STONES AROUND YOU

CONNECTING TO PLACE

Sometimes we can neglect the crystals and stones beneath our feet, yet these are often the ones we have the deepest relationship with. Over the years I have come to notice that different places hold specific energies. My family are from Bath in southwest England, a famous Georgian city depicted in many period dramas. The local limestone, Bath Stone, holds both the vibration of the water found in its Roman baths and the regal energies of Bath's Georgian past.

This month I invite you to explore the energy of the stones and crystals that make up your own local landscape by taking a stroll to tour its monuments, public buildings and ancient structures.

HOW TO USE THEM

Consider taking time to find a local stone circle, monument or a heritage site you are drawn to. Once there, find a spot near the stone(s) where you won't be disturbed. Close your eyes and, if the site allows you to touch the stone, place your dominant hand on it. Alternatively, you can hover your hand over the stone. Take some breaths to stay relaxed.

To deepen the connection, think of three words to describe how the stone feels. Clear your mind and ask the stone for permission to communicate with it. You may hear a yes or feel the positive energy of a yes. If you feel anything else, ask if you can connect again at another time. If you receive a yes, ask the stone to show you some visions in your mind's eye, or to share three words with you. Focus on your breath and allow the messages to come; it might take a few moments. You can then see if the stone will let you ask more questions.

Once finished, leave an offering to say thank you. This could mean placing the stone you used in the window so it can be charged by the sun, or putting it next to a Clear Quartz to amplify its energies. For stones and crystals that remain outside, you might consider placing some flowers, shells or a complementing crystal there. The gift should be natural items that are in keeping with the area, rather than man made.

Additional properties: Grounding, timelines and past life connections

Source: Worldwide

Crystal shape: Raw

Best times to work with this crystal: Mondays, mornings

APRIL ASTROLOGICAL CRYSTAL: **PYRITE**

FINANCIAL PROSPERITY

April, with its Taurus energy, is the time to focus on financial abundance.
Taurus asks us to build our life on strong foundations. Pyrite is the crystal for
financial abundance. During the California Gold Rush it was mistaken for
gold and named Fool's Gold, its connection to prosperity thereby cemented.

HOW TO USE IT

Square shapes represent stability and safety, so to use the properties of Pyrite,
consider creating a square crystal grid to help attract financial stability and
abundance. You will need Pyrite cubes for the centre stone and desire stones,
and optional Clear Quartz points to amplify and direct the energy.

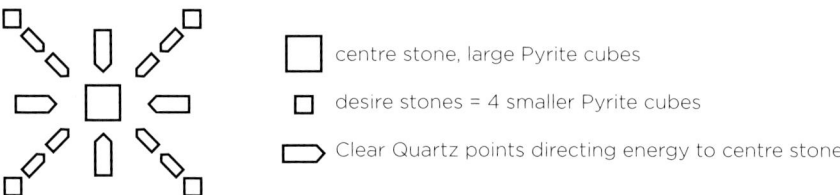

centre stone, large Pyrite cubes

desire stones = 4 smaller Pyrite cubes

Clear Quartz points directing energy to centre stone

1. Start by writing in your journal about exactly where you want more
    abundance in your life. You might even want to note it down on a piece
    of paper that can be placed under the centre stone.

2. While contemplating the outcome you want, start building your grid,
    following the diagram above. Add the crystals in this order: centre stone,
    then four desire stones to create the grid's square shape. Add the Clear
    Quartz points, if using, around the crystals so they are directing the
    energy inwards.

3. To activate the grid, contemplate the outcome you want and touch the
    centre stone with your finger. Touch any of the surrounding stones,
    then go back and touch the centre stone again, then out to the next
    surrounding stone. Keep doing this until you have touched all the stones.

4. Keep your crystal grid intact until the intention has been met, or if your
    desired outcome has changed. Mindfully thank the grid, take it down and
    cleanse the crystals (see page 18).

Additional properties: Supports the divine masculine, abundance and new opportunities

Sources: Peru, Italy and Spain

Crystal shapes: Towers and natural squares

Best times to work with this crystal: Thursday, afternoons

APRIL GODDESS CRYSTAL: **CHAROITE**

SPIRITUAL DISCERNMENT

Sophia loves all humans and wants to guide us all individually and collectively towards something better. She is known as a creator goddess, which means she has unimaginable abilities to create, and can use signs and symbols in the real world to offer us guidance. As the goddess of wisdom, she can step in to support us when we don't have the knowledge or experience to make the best decision.

The purple and black in Charoite, also known Charoite Jade, creates a striking crystal that represents how Sophia wants to work with us. The purples speak of Sophia's ability to offer us spiritual guidance through intuitive messages, while the black brings Sophia's calming, peaceful reassurance, which tells us that she can offer the wisdom to make better decisions.

HOW TO USE IT

Source a Charoite tumble or a piece Charoite jewellery you are attracted to, then keep it for moments when you want to invite in Sophia's wisdom. As you put on the jewellery or put the tumble in your pocket, repeat the following phrase three times: *Through Charoite I invite Sophia's wisdom.*

Remember: since Charoite is a purple crystal, the insights will come in the form of intuitive messages, but the crystal's black tones indicate the physical world Sophia is believed to have created. Therefore, she is likely to communicate with you using physical symbols in the real world. You might see the same road sign multiple times, pick up a message in songs or people's conversations, or notice a repeating number. Wearing a crystal with both purple and the black tones helps us discern intuitive symbols from simple everyday items.

Additional properties: Transforms, heals karma, teaches us to use our intuition as an everyday tool

Source: Russia

Crystal shapes: Jewellery and tumbles

Best times to work with this crystal: Monday, evenings

APRIL CRYSTAL OF THE MONTH:
LAPIS LAZULI

ANCIENT WISDOM

April is a month full of wisdom. Lapis Lazuli is also known for its wise teachings, making it the perfect crystal to reflect the energies of the month. Lapis Lazuli is a metamorphic rock comprising Lazurite, Calcite and Pyrite, and is formed by contact metamorphism, the exposure to extreme pressure and heat through lava. The result is a crystal that has gone through a powerful change, and therefore wants to help us do the same.

The ancient Egyptians loved Lapis Lazuli, and saw it as a facilitator of ancient wisdom. Its deep blue tones resonate with the throat chakra, helping it communicate those teachings so that we can assimilate it into something we can then share with others. Interestingly, some mystics believe the chakras of the ancient Egyptians were higher. The throat chakra was positioned where we know the third eye to be, indicating that they could have potentially communicated with each other telepathically via their third eye.

Today, we can use Lapis Lazuli to expand the amount of intuitive information we receive. Lapis Lazuli is also said to have been around 'before time began', offering us wisdom as well as a direct connection to universal consciousness.

HOW TO USE IT

I strongly recommend purchasing a specimen of Lapis Lazuli that you can wear. This allows the crystal to work with your energies all day, raising your vibration so you can access these higher forms of guidance. Then, each night before you go to bed, place your crystal on your third eye for 10 minutes. Close your eyes and relax. Trust that, over time, this crystal will unlock and enhance your claircognizant abilities or intuitive knowings, opening you up to receive intuitive information.

Additional properties: Telepathy, communication skills and trust

Sources: Afghanistan, Chile, the USA and Russia

Crystal shapes: Jewellery and tumbles

Best times to work with this crystal: Monday, mornings

APRIL ACTIVITIES

1. CONNECTING TO THE EARTH'S CRYSTAL GRIDS

On page 70 I walked you through the process of connecting to stones in your local area and ancient monuments, but it is also possible to connect to the crystals in the ground as well. This practice is a way of honouring Mother Nature herself. It also allows you to draw up the energies of crystals you might not have in your collection, so that you can work with them as well.

USING YOUR EARTH STAR CHAKRA

1. Lie down comfortably. You may want a pillow behind your head and a bolster or pillow underneath your knees.

2. Focus on your breath, lengthening it to help you relax.

3. Imagine a large pearlescent sphere about 1 metre (3 feet) beneath your feet. This is your earth star chakra, and it operates as our connection to Mother Nature. This energy also helps us connect to the Earth's crystal grids.

4. Imagine that you are travelling into the ground, going past different layers of crystals. After a while you find the crystal you want to work with. Imagine drawing that crystal's colour up and into your earth star chakra. See the chakra change to that colour.

5. Now imagine that colour coming up and surrounding you. Stay there for about 5 minutes. When you have finished, visualize the energy going back into the earth star chakra, and then back into the ground. Open your eyes.

2. ROSE QUARTZ SELF-CARE RITUAL

Aries has most likely had you rushing around, putting plans into place and implementing new projects. Now it's time to check in and see what you need so that you don't drain your energy reserves. This month's connection to Taurus and, subsequently, Venus, also has us considering how we can support our body, offering it some self-care where needed.

MASSAGE AND BATHING

In honour of the planet of love, I'm going to suggest working with Rose Quartz when creating a self-care routine for yourself. There are a variety of things you can do. It is possible to purchase massage crystal tools for the body and face, such as a roller or *gua sha* (a smooth massage stone from traditional Chinese medicine). Both work by alleviating fluid build-up in the body and relaxing the muscles.

I introduced you to crystal elixirs on page 46. We can use those principles and take them a step further by putting crystals next to the bath water to make a crystal elixir bath, allowing us to absorb the energies. Therapeutic baths are a traditional healing and cleansing practice that go all the way back to ancient Babylonia. The Egyptians and Greeks loved their ritual baths and would add herbs, flowers, milks and salts to the water.

Here is a self-care bath recipe fit for the goddess Venus herself:

1. Place your Rose Quartz crystal next to the bathtub.
2. Add two drops of rose water, a pot of rosehip tea (strained) and a handful of Epsom salts to the bath water.
3. Float pink flowers and rose petals on the water.

May

SPRING ENDS: **COMMUNITY, INSPIRATION & RELATIONSHIPS**

May sees nature in full bloom. Wildflowers flourish, honey bees collect pollen and spring fruits and vegetables ripen, ready to be picked.

After April's slower pace, May coaxes us to go quicker again. Inspired by Taurus, we start the month checking, then refining, our plans. Then Gemini takes over, offering inspiration and new ideas to improve our goals. However, this time our focus has changed. Alongside our summer projects, now our main priority is getting outside and spending time with our loved ones.

This coincides with the Beltane or May Day celebrations. As the farmers start to see the fruits of their labour flourishing in the fields, the community comes together to honour Earth's fertility, hoping that their gratitude will later turn into a plentiful harvest. This is the last month of Maiden energy, and the bringing together of the community marks a time when people often met and fell in love: romance and new relationships is a theme for May. The word May derives from the Greek goddess Maia, goddess of fertility and growth.

This month we connect to the ancient Egyptian goddess Isis, and I suggest crystals to help you strengthen your intuition, connect with your body and attract love.

May facts
Colours: Pink, green, yellow and orange
Crystal shapes: Spheres, hearts and clusters
Seasonal chakra energies: Heart, sacral, crown and third eye chakras
Animals: Penguins, meerkats and giraffes
Seasonal shrine/altar: Grass, wild flowers, maypole ribbons, flower headdress, May baskets
Intention setting/affirmation: *I welcome more love and joy into my life*
May birthstone (traditional and modern): Emerald
Birthstone properties: Understanding lessons of the heart. Growth and nature

MAY SABBATH CELEBRATION: **BELTANE**

After a busy few months working on the land, preparing the soil and sowing seeds, farmers are now caring for their growing crops. The Celtic festival Beltane, also known as May Day, is the moment the communities come together on 1 May to honour Mother Nature. People typically give thanks and manifest a bountiful harvest in the coming months.

The second cross-quarter fire celebration, Beltane honours spring's connection to the Earth element. For that reason, people collected the ash from the fires and used it to bless the crops and their homes.

During a Beltane bonfire ritual, community members jumped or ran through the fire to receive protection. Some jumped to safeguard their crops, others to guarantee a safe birth or safe travel. Children were carried through. The eldest female in the community walked around the fire three times to ensure illness would not come to anyone. Young women seeking love were traditionally the last to run through as the fires calmed and started to burn out. On this day the cattle would also be moved from winter to summer pastures, passing the bonfires so they could also receive protection.

A magical time, Beltane has a close connection to Samhain (also known as Halloween). On both days, the veil between the living and the dead is believed to be at its thinnest. While Samhain concentrates on those who have passed away, Beltane focuses on the souls who want to be reborn.

New couples coming together at this time of year traditionally spent May Day night together in the woods. Winter babies born in December and January are believed to have strong psychic abilities and a connection to the faye folk (fairies).

CRYSTALS TO DISPLAY FOR BELTANE

Honouring growth: Emerald and Cherry Blossom Agate

Finding love: Rose Quartz and Pink Kunzite

Fertility: Garnet and Ruby

MAY ASTROLOGY
& FULL MOON

We begin May, inspired by Taurus, evaluating the areas of our life that need stronger foundations so that our future achievements are long-lasting. We then enter Gemini season from 21 May.

By now the lighter, warmer days have moved our attention away from our goals and towards outdoor pursuits. We prioritize having fun, being out in nature and seeing our wider community. May asks us to keep working on our projects when we can, but it knows that time outside gives us space for inspiration to come through.

Gemini is an Air sign, so it likes to contemplate and think things over. However, remember that Gemini's symbol is the twins, so it can look in two directions, offer opposing views and change its mind regularly. Therefore, this is a month to note down good ideas you have, but wait until Cancer season next month to decide which to pursue.

The symbol of the twins brings up the topic of duality this month: good and bad, right and wrong, light and dark. This might show up as self-sabotage, so if you progress your goals and feel negative mind chatter coming to the surface, carry your Pink Lithium Quartz crystal (see page 28) with you to help calm those thoughts. You can also keep a notebook to record regular negative thoughts so you know what to reflect on and heal next January.

MAY'S FULL MOON

This full moon is traditionally known as the Flower Moon because of the month's abundance of flora and fauna. May's full moon has numerous other names, most of which refer to fertility and growth, such as the Hare Moon and the Milk Moon.

MAY GODDESS: ISIS

NAME AND MEANING

This ancient Egyptian goddess superseded her origins to be worshipped in ancient Rome, then in pagan teaching, and is still honoured worldwide today. Isis, also known as the Goddess of Ten Thousand Names, is the goddess of love, fertility, reassurance, magic, healing, the skies, marriage and motherhood. Her name means 'throne'.

CHARACTERISTICS

Isis and husband Osiris were the first rulers of the world, and Isis is said to have given mankind a remarkable evolutionary boost and formed the first civilizations. She is depicted with outstretched wings, protecting all of Egypt.

Osiris's jealous brother Seth wanted to rule, so he killed Osiris and spread pieces of his body all over Egypt. Isis brought Osiris back from the dead and he became King of the Underworld. Meanwhile, Isis had his baby, Horus, and together he and Isis hid until Horus was old enough to take back the throne. This magical conception, as well as Isis's unwavering loyalty to her son, led to her reputation as a fertility goddess and perfect wife. She is also known for her unwavering determination and magical ability to overcome obstacles.

Isis also appears in the constellation Sirius, the brightest star in the sky. Each year, when that star appears in Egypt, it marks the start of the great Nile flood and Egypt's agricultural season, and welcomes in the new year. Some cite the great flood as Isis's tears for the loss of her husband.

TEACHING

We can call on Isis to protect our family and add inspiration to our dreams so they can grow. She can inspire us to work the impossible when life is especially complicated.

SYMBOLS OF ISIS

Wings, ankh, knot of Isis, cows and the solar disks

MAY SABBATH CRYSTAL:
PEACH MOONSTONE

CYCLES, THE MOON AND FEMALE FERTILITY

As their name indicates, all crystals in the Moonstone family resonate with the moon and her cycles. That means they are good at connecting us to the cycles of nature and supporting us through phases of our life. That could be the start of a new project, seeing ongoing projects through to the end or transitioning from the end of something and onto something new.

Peach Moonstone connects us to female cycles and has been historically used to help women through the stages of pregnancy, birth and motherhood. With this crystal's connection to the monthly and seasonal cycles, as well as female fertility, it is the perfect May crystal.

CONNECTING TO THE WOMB

The womb is a sacred space for all women, whether they plan to have a family or not. An area of sacred growth, it is also where emotional wounds can reside. Like the heart and stomach, the womb can offer profound intuitive guidance.

On a full or new moon, place a Peach Moonstone on the pubic bone for 10 minutes to help relieve any old emotions that might be there. At the same time, this crystal will strengthen that area, supporting both fertility, female health and the growth of new things in our life.

Once old energies have been released, you can use your crystal to gain intuitive insights. To do this, place your crystal on the pubic bone, relax for a few moments, take a slow inhale, then ask the crystal a question and allow the answers to come on the exhale.

Additional properties: The growth of new projects, transition and skills of clairtangency (the ability to psychically receive information through touch)

Sources: Sri Lanka, India, Brazil, Madagascar, the USA and Myanmar

Crystal shapes: Palms, tumbles and worry stones

Best times to work with this crystal: Friday, mornings

MAY ASTROLOGICAL CRYSTAL:
BLUE KYANITE

INTUITIVE COMMUNICATION

Gemini's symbol of the twins hints at its deep contemplation and its strong intuitive abilities. Each and everyone is likely to receive intuitive guidance at this time of the year. Gemini's planet is Mercury, the planet of communication, which advances its communication abilities even further.

All blue crystals are connected to the throat chakra, and therefore support communication. Within that, each type of crystal focuses on different aspects of communication – and Blue Kyanite enhances intuitive communication.

All psychic messages come into the body through the crown chakra and turn into guidance received by the third eye. If that guidance is received as thoughts or ideas, we might not realize its importance, but Blue Kyanite reinforces the connection between the third eye and throat chakra, which helps us act on the guidance, rather than dismissing it. If required, it then helps us interpret the message into something we can communicate to others.

HOW TO USE IT

Blue Kyanite grows in a natural rod-like shape that allows it to channel its energies. Source yourself a small specimen that you can place over your lips so that one point is facing your nose. Clear Quartz and Kyanite often form together, and if you are new to intuitive work, consider starting with a specimen that includes Quartz. This will amplify Kyanite's energies and help you pick up on the messages more easily.

Lie down and position your crystal over your mouth for 20 minutes a day for one week. After the week has finished, start doing this process once a week for another month. After a few days it should increase the amount of times you wonder *was that my intuition?* after receiving potential intuitive guidance. This seems to be how our intuition gets us to notice its messages – so when you think that thought, the answer is always yes!

You can repeat this technique later on in the year if you find that you have not received any intuitive guidance for a while, as that indicates that the connection needs strengthening again.

Additional properties: Enhances the throat chakra, aids channelling, light language activation

Sources: Brazil, Australia and Zimbabwe

Crystal shapes: Raw, tumbles and points

Best times to work with this crystal: Monday, mornings

MAY GODDESS CRYSTAL:
LIBYAN DESERT GLASS

UNIVERSAL WISDOM

Libyan Desert Glass is also called Libyan Gold Tektite. Tektites are a form of natural glass formed from terrestrial debris that is released from a meteorite on impact. This means it can bring us new, unexplored 'alien' concepts and ideas.

Scientists do not understand how Libyan Desert Glass was formed, but they do acknowledge that whatever happened, its creation was intense and dramatic. The result is a crystal with the ability to transform us rapidly.

If you don't know where to start with your current goals or you're hitting a road block, you can ask the cosmic goddess Isis to use her connections to Sirius to bring in new ideas. Tektites are perfect for this kind of work because they hold that galactic energy.

Libyan Desert Glass can also give you access to your Akashic records. These are the register of all the past lives, lessons, skills and karma our souls have collected. It allows us to bring the wisdom of those lives forward so we can embody them in this lifetime.

HOW TO USE IT

Display a statue or photo of Isis in a prominent spot in your home. Before you start asking anything, it is a good idea to offer all deities a gift. Red Jasper was regularly used in jewellery associated with Isis, so you could position a Red Jasper crystal next to her image as an offering. Now place your Libyan Desert Glass specimen by her image, leaving it there overnight – the intention is that Isis will impart her wisdom into the crystal. Keep it with you the following day. Place it next to her again before you go to bed. Over time, new pathways will start revealing themselves to you.

Additional properties: Soul purpose support, explores new concepts, promotes abstract thinking

Sources: Libya and Egypt

Crystal shapes: Raw and jewellery

Best times to work with this crystal: Wednesday, early afternoons

MAY CRYSTAL OF THE MONTH: EMERALD

GROWTH

Emerald, the birthstone of May, is closely connected to the history of man, and some of the best-known crystal myths and legends are associated with it. This includes Cleopatra's love affair with Emerald, which motivated her to declare ownership of the first Emerald mine in ancient Egypt, the Egyptian god Thoth's Emerald tablets and the myth that states if you place a specimen under the tongue, you can tell the future.

Emerald, a type of Green Beryl, is a precious stone, one of the most valuable and desirable crystals in the world. When it is faceted in jewellery and held up into the light, it is believed to hold the purest green light frequency, giving it the most potent and healing properties of all green crystals. In this case it offers us the ability to find the underlying cause of heartbreak in our lives and the wisdom we need to heal and then make different decisions in the future.

HOW TO USE IT

If May is all about finding our ideal partner, it makes perfect sense to first work with Emerald to heal the past so that we can make better relationship decisions in the future. This involves wearing an Emerald necklace so that the crystal is over your heart, allowing it to release any old energies.

At the same time, it will also guide you to make different decisions in the future, so if you find yourself drawn to people who are not normally 'your type', go with it, that could be Emerald guiding you in a new direction.

Additional properties: Balances the heart chakra, releases grief, cultivates better relationships

Sources: Brazil, Egypt and Colombia

Crystal shapes: Raw and jewellery

Best times to work with this crystal: Friday, late mornings or before bedtime

MAY ACTIVITIES

1. CRYSTAL LOVE SPELL BOTTLE

In the past, when our ancestors needed support, they would have visited the wise woman at the edge of the community who would work her magic to create something like a spell bottle they could carry with them. Today we call spells 'manifestation work'.

Spell bottles contain items that symbolize what you would like to attract or hold the energies of it, so it's a step on from carrying a crystal with you or making a vision board.

MAKE A SPELL TO MANIFEST A PARTNER

You will need a miniature brown glass bottle with cork, paper and pen, matches, an incense stick, Emerald and Rose Quartz crystal chips, rose petals and a green candle.

1. Clean the bottle (if necessary). On a piece of paper small enough to fit inside the bottle, list the attributes you want your partner to have. While you focus on the outcome you want, light the incense and use it to cleanse the inside of the bottle, roll or fold the list and drop it inside, add the crystal chips and rose petals and seal the bottle with the cork. Light the green candle and allow the wax to fall over the cork, sealing it in place.

2. Hold the bottle in your hands, close your eyes and imagine a green light surrounding it. Say to yourself three times: *May my words be truth.* Then carry your bottle with you in your bag, or keep it near where you sleep.

2. FERTILITY CRYSTALS

Prayer, blessings and rituals focused on fertility have always been connected to spring, and we can embrace these energies today to support our own fertility journey.

USING CRYSTALS TO AID CONCEPTION

Historically, our ancestors placed root chakra crystals like Garnet and Ruby under the bed for fertility support. I would take that a step further, and suggest placing a red crystal on each corner of the bed to create a grid that supports conception.

If you are pregnant, you can display a Peach Moonstone crystal on your bedside table to support you and your baby.

For those finding it hard to conceive, the self-care steps on page 174 would be a loving support for you and your body. I have also suggested some crystals below and how they might help you. Once you have pinpointed the best one for you, place it over the relevant chakra for 3 minutes every other day.

ROOT CHAKRA

Garnet: Male fertility

Pyrite: Fears about whether you will be a good parent

SACRAL CHAKRA

Peach Moonstone: Female fertility

Ocean Jasper: Irregular cycles

Tangerine Quartz: Sexual trauma

June

SUMMER BEGINS: **EMOTIONS, GOALS & EMPOWERMENT**

Summer is the Fire element season, motivating us to work and achieve, but also to play hard, travel and create memories. While our diaries fill with activities, Mother Nature has moved from the Maiden to the Mother goddess, caring for the baby animals and holding space for us to play and enjoy life.

Mid-month, we celebrate the summer solstice. There are light codes in the sun's rays which, when passed through the body, offer insights and encourage action. This month you will find motivation to begin working on your goals, or, if you are already working hard, to increase momentum. The Hindu goddess Lakshmi, will support you in this.

Through the summer months our solar plexus is fully active. It is connected to our digestive system, and we seek out cooling foods to contrast the hot days. You can also carry blue crystals to cool the body. Our increased vitality also comes from the solar plexus, as does the heightened need to explore, travel and achieve our goals. When we are working from the solar plexus its correlating fears can appear: imposter syndrome, lack of confidence and related procrastination. I provide a crystal elixir recipe on page 106 which will relieve these thought patterns and help you overcome these worries.

June facts
Colours: Yellow and green
Crystal shapes: Towers, points, spheres and floral shapes
Seasonal chakra energy: Solar plexus chakra
Animals: Bees, ladybirds and eagles
Seasonal shrine/altar: Oak, honey, fresh herbs, mead, yellow candles, symbols of the sun and fairies
Intention setting/affirmation: *I have boundless energy*
June birthstone (traditional): Pearl
Birthstone properties: Reflection, transformation, tranquillity
June birthstone (modern): Alexandrite
Birthstone properties: Adaptability, insight, drive

JUNE SABBATH CELEBRATION: LITHA

Litha celebrations run from 19–23 June, the main focus being the summer solstice, the longest day of the year. This normally falls on 21 June, but depending on the year and time zone it can occur on the dates either side.

On this day, the northern hemisphere experiences its maximum tilt towards the sun, creating the longest daylight and shortest night. Because this time is viewed as the middle of summer, it is also often referred to as Midsummer.

The word 'solstice' is derived from the Latin *sol* ('sun') and *sistere* ('to stand still'). On this date the sun appears to stand still, pause and then reverse direction. Our ancestors lit bonfires and midsummer beacons to help the sun progress through the sky. They embraced the longest day, making the most of the light and offering gratitude for life, fertility and the abundance all around them, in hope that their admiration would make the gods happy and give them plentiful crops to see them through the coming winter months.

Litha is a wheel of the year celebration that marks this auspicious date. Symbols of the sun would have been everywhere. It also marks the moment when the Oak King, a symbol of the sun, reaches the peak of his power. His reign is coming to an end and he hands over power to the Holly King, overseer of the darker nights.

This was seen as one of the most magical times of the year; people believed that fairies came into the world on this date to offer blessings. Herbs and flowers needed for magical work and blessings were collected at midsummer so they could harness the fairies' magic.

CRYSTALS TO DISPLAY FOR LITHA

Symbols of the sun: Sunstone, Fire Quartz, Amber and Gold

Summer celebrations: Strawberry Quartz, Honey Calcite and Laughing Stone

Fairy crystals: Spirit Quartz

Fertility: Peach Moonstone, Garnet and Ruby

JUNE ASTROLOGY
& FULL MOON

June begins in Gemini's energies, meaning that moments of ideas and inspiration are still coming. Then, on 21 June, around summer solstice, we move into Cancer season.

The crab perfectly reflects Cancerian traits and the lessons it wants to offer us. The crab shell represents the barrier or mask we can put up to hide ourselves from the world. It is a Water sign, indicating the hidden aspects are going to be emotional, such as fears, complex emotions or perceived weaknesses. And although the summer sun asks us to progress our biggest goals, Cancer can introduce self-sabotage and feelings of imposter syndrome.

A cardinal sign, Cancer has an air of authority and commands our attention. Like the pinch of a crab, it will make us aware of the emotions that need healing. Cancer's planet is the moon, so both sign and planet are connected to the Water elements. This means emotions can run high. While the Water sign Scorpio relates to storm clouds and sudden downpours of emotions, and Pisces relates to the river and progression up or downstream, Cancer asks us to look at the vast ocean of our emotions. Like the ebb and flow of the tide, our aim this month is to calm the waves and heal what is hidden in the depths of the sea.

Finally, Water star signs have a great gift of intuition, and Cancer uses our gut instinct and clairsentient skills (how we feel about something) to offer us intuitive insights. Now is the time to take the list of ideas the Gemini season offered you and consider how you feel about each one. Cancer will help you understand which ones you should progress.

JUNE'S FULL MOON

This full moon is known as the Strawberry Moon because it's the time of the strawberry harvest. The two additional names for this full moon are the Hot Moon and Rose Moon. This full moon shines a flashlight on what is going on: if you notice that specific emotions are building right now, take it as an indication of what needs healing.

JUNE GODDESS: LAKSHMI

NAME AND MEANING

The Hindu goddess Lakshmi's name derives from
the Sanskrit word for 'goal', and she is the goddess of
wealth, fortune and prosperity, power, love, abundance
and beauty. She is our ideal companion for the summer.

CHARACTERISTICS

Lakshmi is the daughter of the Hindu mother goddess Durga. She teaches us
to transcend the material world and instead explore true abundance in our
spiritual growth. She is depicted sitting on a lotus flower, which is symbolic of
the path Lakshmi asks us to take: this is a flower that starts life in the mud, then
grows through murky waters, before finally reaching the surface to blossom.

According to legend, Lakshmi came into the world as an adult female, floating
through the sea on a pink lotus. She is often portrayed as a fertility goddess, the
feminine counterpart to Lord Vishnu. Her beauty, purity and grace means that
she is beloved in Hindu culture and honoured in the annual Diwali festival.

TEACHING

Lakshmi's teachings speak of the four transformations people can make
through reincarnation. Each one is represented by one of her four arms.

Dharma: An ethical and moral life

Artha: Wealth and material enjoyment

Kama: Love and emotional fulfilment

Moksha: Self-knowledge and liberation

Lakshmi appears in eight divine forms, each representing a different area of
prosperity. Working with Lakshmi offers profound guidance and reminds us
to make sure our goals are for our highest good.

SYMBOLS OF LAKSHMI

Pink lotus, golden coins, the colours red and gold

JUNE SABBATH CRYSTAL:
AMBER

AUTHENTICITY

The summer solstice is the perfect time for you to connect more deeply to the solar plexus chakra. Known as the home of your 'I am', this is our own central sun. It represents our soul's purpose and who we have come here to be, and includes our personality, motivations and interests.

Our 'I am' is an expression of our ego. While an over-inflated ego is not healthy, the ego is the part of us that wants to show up and achieve things, so it is key to our progression.

If you feel you don't have the confidence or motivation to strive towards your bigger goals, or if you start a project but quickly run out of steam, you can work with yellow crystals to enhance the energy of the solar plexus, which should give you the nudge you need.

Amber is fossilized tree resin that forms into yellow, butterscotch and brown specimens, often with insects encased inside. In my experience its sunny tones gives our solar plexus the most powerful injection of vitality.

CENTRAL SUN VISUALIZATION

This is a visualization that uses Amber to expand our inner sun, helping us embody our soul's purpose and raise our confidence levels. It shows us how to be confident in our own skin.

1. Sit down and, using your non-dominant hand, place your Amber crystal against your solar plexus. Close your eyes and relax.

2. Where your solar plexus is, visualize a small sun.

3. In your mind's eye, list all the things you would do and achieve if you had the confidence, and see those words sitting in that sun.

4. Imagine that the sun increases in size until it completely surrounds you.

5. Stay there for a while, and then open your eyes.

Additional properties: Confidence, creativity and self-expression

Sources: Germany, Poland and Russia

Crystal shapes: Raw and jewellery

Best times to work with this crystal: Sunday, early afternoons

JUNE ASTROLOGICAL CRYSTAL: UNAKITE

EMOTIONAL RELEASE

This month, Cancer asks us to be brave and heal hidden, suppressed emotions. Unakite is the perfect crystal for this kind of work because it helps release the energy around those emotions, so we don't necessarily have to talk through traumatic events before the energy can start to release.

Unakite is a dense crystal, which symbolizes how it can be used as a tool in the physical world. A metamorphic rock, it was originally made up of multiple crystals (Pink Orthoclase Feldspar, Epidote and White Quartz) that have been forged together and, like all metamorphic rocks, it is good at supporting change. They work together to access both the green and pink frequencies connected to the heart chakra.

Pink Orthoclase Feldspar: Holds a higher heart chakra healing frequency that raises our vibration above the emotions that need healing.

Epidote: Moves us past old, stuck emotions and helps us navigate who we are without those emotions holding us back.

White Quartz: Adds clarity to the situation, so even if we cannot name the emotion, we can see how much better we feel once it has gone.

HOW TO USE IT

The easiest way to use a crystal to release emotions is to keep it nearby as you sleep. While we sleep, our conscious mind is resting, and this makes our subconscious, which holds suppressed emotions, more accessible.

I suggest displaying a Unakite specimen by your bedside, either in a pyramid shape to represent the two extremes of emotions and balance the mind, or in a heart shape for compassion and love.

Additional properties: Balances emotions, lifts our spirits, forges a connection between the head and the heart

Sources: The USA, Zimbabwe and Switzerland

Crystal shapes: Hearts, pyramids, jewellery and double-terminated points

Best times to work with this crystal: Friday, late mornings or before bed

JUNE GODDESS CRYSTAL:
PERIDOT

POSITIVITY

Known as the crystal of positivity, Peridot's message is to keep going: anything is possible! A once-favoured gemstone, Peridot (sometimes called Olivine) is often overlooked these days, but once you read about its properties, you will understand why it deserves to be better known.

Olive-green in colour, Peridot is a combination of the heart chakra's green and the yellow of the solar plexus. Historically, olive represented the divine feminine. It honoured our soul's purpose and the lessons our heart has to offer. If we operate from the heart, not the head, the outcome will always be more rewarding.

You might remember that Lakshmi's lessons were to look past the material world and aspire to higher forms of wealth and prosperity. Closely connected to Peridot, she knows that this crystal can help us understand how that relates to our life plan.

HOW TO USE IT

This is a crystal that always looks best when faceted into jewellery, so I recommend finding a pendant that can be worn on a long chain that allows it to fall between the heart and solar plexus. Wear it when you're deep into a project and need some motivation, or create an elixir from this crystal when you need an emotional pick-me-up.

Additional properties: Motivation, female empowerment and focusing on the positive

Sources: Australia, Pakistan, China, Egypt and Sri Lanka

Crystal shapes: Jewellery and chips

Best times to work with this crystal: Sunday, afternoons

JUNE CRYSTAL OF THE MONTH: DENDRITIC OPAL

MAGIC

Dendritic Opal combines all the properties June has to offer: growth, nature, the emotional depths and wisdoms of the Cancer season and an unlocking of past-life memories that ancient ceremonies, such as the summer solstice, invoke in us.

The high water content of all Opals represents their ability to support our emotional body, while the dark dendrites (the tree-like branches inside the crystal) symbolize growth and development. This is a powerful property in any white crystal, as it represents learnings relating to the crown chakra and our spirituality. When this is found in a dense crystal like Dendritic Opal, bringing it into your life will expand your spiritual development.

This is where its alternative name comes in. Inspired by the famous wizard, Merlin, Merlinite helps you to understand what your magical gifts are. These could be psychic skills, healing abilities, manifesting skills or creative pursuits.

CLAIMING YOUR MAGIC CRYSTAL LAYOUT

To bring the knowledge and wisdom of your magical skills into your awareness, we need to bring the information into the body through the crown chakra. To do this, place a large Black Tourmaline crystal under your feet, then lie down. Place a good-sized Clear Quartz point above the head, with the point facing you. Put a Dendritic Opal stone in each hand and lie there for 20 minutes.

Once the time is up, place your Dendritic Opal crystals over your heart. On an inhale, ask both crystals to show you an image or word that represents your gift. Then, on a slow exhale, see what information comes.

Additional properties: Spiritual awakening, connects the physical and energic bodies and develops your intuitive skills

Source: The USA

Crystal shapes: Palms or hearts

Best times to work with this crystal: Monday, late mornings

JUNE ACTIVITIES

1. CRYSTAL-INFUSED SOLAR WATER

The height of the summer sun also marks the time when we'll be deep into progressing our yearly goals. As we create something new, limiting beliefs or blocks can start playing out, slowing or stopping our progression.

Your ego likes your normal routine because it's safe. When we work on our goals, our ego doesn't know if these new behaviours will keep us safe, so it uses negative mind chatter to try and stop us doing anything new.

Logically, we know the changes are good. Therefore, we need to quieten the egoic mind so we can keep going. Crystals connected to the energies of the sun can help us to do this, and solar water can help you take on their energies.

SUMMER SOLSTICE SOLAR WATER

Solar water is an elixir that has been outside in the morning sun, so it can be charged by the sun's positive energy. I suggest adding a Sunstone next to the water to amplify the sun's energy, as this will give you the confidence to succeed.

When fear starts holding you back from the direction you want to take, you can use this elixir to raise your vibration high enough that those negative thoughts dissipate. To create an extra-special elixir, try to do this on the morning of the summer solstice.

You'll need a Sunstone, drinking water (ideally distilled), a glass jug, some muslin (or cheesecloth) and some ribbon.

1. Rinse your crystal under water.
2. Pour distilled water into the jug. Cover the jug using muslin tied in place with the ribbon.
3. Leave the jug outside in the morning sunlight with the crystal placed next to it.
4. Remove the muslin and drink the water.

2. SUMMER SOLSTICE DAY RETREAT

The summer solstice is the perfect time for honouring the sun's cycle, the twenty-four-hour cycle of day and night. The sun has a short cycle allowing fast progression. The focus is on understanding what success means to you, goal setting, activating your creativity and vitality, and empowering yourself. So, when you are reaching a new stage in a project, or have a problem you want to overcome quickly, set a day aside to follow the sun's cycle.

SUNOLOGY RETREAT SCHEDULE

Sunrise: Start the day focusing on new beginnings. Cleanse your space, write down your goals for the day and prepare anything you will need. Consider making some solar water, as detailed on the previous page.

Morning: As the sun gets higher, your positivity levels increase, making this the ideal time to problem-solve and expand your ideas. Keep a yellow crystal with you to enhance the morning's positivity and make notes of any questions to ask your intuition later.

Midday: At the sun's highest point, its magic is at its fullest and you can seek intuitive guidance through its rays. Go outside, retrieve your solar water and journal about the questions you collected in the morning. Read a question and then, without thinking, allow yourself to intuitively start writing the answer. Purple crystals will help you connect to your intuition.

Afternoon: The afternoon's energy is more practical. This is the time to use the momentum of the day to put your ideas into action. If you need to contact people, add jobs to your diary or start researching, do it now. Keep a calming blue crystal by your side.

Sunset: As the sun disappears, work on anything that might be holding you back from your goals. Perhaps you have limiting beliefs, or you are repeating the same patterns. If this is the case, release them into the sunset with Black Moonstone (see page 34).

July

SUMMER: **JOY, LIBERATION & PURPOSE**

There is no Sabbath this month. Historically, this was a time when whole communities would be in the fields working the land, harvesting the cereal crops and collecting the hay. Everyone came together, as they knew this produce would see them through the winter months.

In a dramatic contrast today, machinery takes on a majority of this work. For everyone else, July is all about having fun. Children's summer holidays have started or are fast approaching, and adults might start the month working hard, creating space so they can later join their families on summer vacation.

This is the time to let that inner child shine, have fun and play. The Fire element motivates you to be adventurous and do things you might not normally do.

When work does call, particularly as we enter Leo season, embrace July as a time to explore and pursue your most audacious goals. July is the zenith of creative energy, so try and ride on its coattails. The height of summer is full of heat, warm breezes and long, sunny days that push us forward.

Leo season reminds us not to accept second best. Instead, it asks us to become the best version of ourselves. This month I will show you how to work with crystals to embrace all this potential and highlight how the ancient Egyptian goddess Hathor and the heightened energies of Lion's Gate (see page 120) can make this your most successful year to date.

July facts
Colours: Yellow and gold
Crystal shapes: Tumbles, clusters, merkabas and skulls
Seasonal chakra energies: Heart, solar plexus, sacral and root chakras
Animals: Lions, snakes and butterflies
Seasonal shrine/altar: Seasonal music, honey, wine, daisies and sunflowers
Intention setting/affirmation: *I align with my soul's purpose*
July birthstone (traditional and modern): Ruby
Birthstone properties: Courage, passion and strength

JULY ELEMENT:
FIRE & SUMMER

From the start of spring onwards, the sun's heat increases incrementally. The shift is subtle, so we may not register the change until suddenly we are more awake. We're not getting out of bed in the dark any more, and the heat of the midsummer sun adds an instant flush of warmth to our skin.

It is no surprise that Fire is the summer's element. Of all the elements, it is the most elusive; whereas Air, Earth and Water are all around us, to work with Fire we first have to create it.

The Sabbaths all honour Fire in their own way. In spring and autumn, it's used to cleanse and protect. During midwinter, it lights the way in the dark. Now in the summer, it's used as an offering to help the sun shine bright for as long as possible. Perhaps such a great significance is put on this element because its discovery safeguarded us from the cold and allowed us to cook our food.

Each of the four fire festivals take on the properties of that season's element. Lanterns are lit to guide spirits (Water, due to the emotional aspect), ash is collected for offerings (Earth), people run through the flames for protection (Fire) and smoke is created to cleanse (Air).

In the heat of the summer season, July marks the beginning of Lion's Gate (see page 120). This is an important time when the heat of the sun has risen enough that the energies of our planet slightly shift, allowing in greater intuitive guidance, specifically relating to our soul's purpose.

CRYSTALS FOR SUMMER'S FIRE ELEMENT

Passion: Fire Quartz

Creativity: Orange Calcite

Goal setting: Citrine

Positivity: Peridot

Energy: Tiger's Eye

JULY ASTROLOGY
& FULL MOON

We step into July with Cancer still asking us to explore any hidden emotions that might be holding us back from success. Once these have been released, here comes Leo season on 23 July.

Arguably the most impactful of the Fire signs, Leo energy wants to impress you. Imagine the symbol of the lion standing proud, appearing strong and owning its power. Leo season asks all of us how we can replicate this in our own unique way. You will be asked to find out where in your life you are settling for second best, and if you can change that. Once Leo season starts, consider journalling on the following question: *What would I do if I could not fail?* Then witness how Leo helps you to achieve it.

Leo season can be a whirlwind for some. You might find opportunities being taken away from you, but quickly replaced by better ones. This is a month to carry Red Jasper to give you courage and Sodalite to help you trust that better opportunities will be coming.

In contrast, Leo can sometime stop us progressing our plans altogether, out of fear of failure. In those cases, you can embrace the other common July weather, thunder, and work with Fulgurite to intentionally dispel any such thoughts. Fulgurite is a crystal formed when thunder hits sand. It can make sudden changes, similar to a lightning strike, so if you display it on your desk, it can quickly remove any fears that come along.

JULY'S FULL MOON

The Thunder Moon is one name for this full moon, as is Hay Moon and Buck Moon, which refers to the seasonal growth of the buck deer's antlers. At this time of year, the antlers of the male deer have reached peak growth.

JULY GODDESS: HATHOR

NAME AND MEANING

Hathor is thought to have been the feminine counterpart
to the Egyptian sun god Ra, the most important of all
Egyptian gods. This is why you will often see her wearing
a headdress depicting the sun disc, on either side of which
are two cow's antlers that symbolize her connection to
motherhood, nourishment and fertility. She was loved
by women, who prayed to her to protect their young families and
support them during childbirth. Protector of women, she is goddess
of the skies, gemstones, love, fatality, beauty, pleasure and dancing.

CHARACTERISTICS

Seen as the body which all souls reside within, Hathor also oversees all
the pleasures of the physical world. She wants and invites people to enjoy
new experiences, so as you enjoy the summer months, invite her to make
the experiences even better.

It was also believed that Hathor occasionally took on seven different forms,
known as the Seven Hathors, possibly connected to the seven stars that make
up the constellation of the Pleiades, and which indicate Hathor's additional
abilities as a sky goddess.

TEACHING

All seven forms of Hathor were associated with fate and fortune, each form
instinctively knowing the achievements and potential of a newborn's life.
People travelled great distances to receive a reading from one of Hathor's
priests. On page 116 we explore how Hathor's intuitive guidance and work
as the crystal goddess can support us.

SYMBOLS OF HATHOR

Cows, the sun, hand mirrors and gemstones

JULY FIRE ELEMENT CRYSTAL:
FIRE OPAL

TRANSMUTING EMOTIONS

Most people love Fire Opal – I see people navigate towards it all the time. Normally only found in small specimens, this crystal's energy dramatically outperforms its size.

Opals are waxy to the touch, and all hold water inside of them. That means they cannot be displayed in direct sunlight because the heat could dry them out, causing cracks in the crystals – an interesting issue for a crystal with 'fire' in its name. Nevertheless, it is an anomaly that helps us understand its main property.

An Opal's connection to water implies it can support our emotional body. A dry, cracked Opal represents the damage negative emotions have on us, but what specific emotions are we referring to? Fire element emotions are connected to the solar plexus and include: procrastination, shyness and confidence issues, including imposter syndrome.

The multicoloured rainbow flash of a Fire Opal is what makes this crystal so special. Because it has all the colours inside it, it can heal any aspect of our life where these issues might be playing out.

FIRE OPAL BODY LAYOUT

When these negative Fire emotions are showing up in your life, source a raw Fire Opal to do the following. Lie down, then place your crystal over each chakra for 3 minutes, starting at the root chakra. When you get to the crown, start working back down the body again, this time leaving it in place for only 1–2 minutes. At each stage, imagine the Fire element burning away anything that's blocking that chakra.

Additional properties: Motivation, igniting passion and balancing the chakras

Sources: Mexico, Brazil, Australia and Ethiopia

Crystal shapes: Raw and jewellery

Best times to work with this crystal: Saturday, afternoons

JULY ASTROLOGICAL CRYSTAL: VANADINITE

WILLPOWER

To complement the willpower and sheer determination of Leo season, we need a crystal whose aspirations and ability to bring things into our lives can match that tenacity. Vanadinite is said to be able to put 'spirit into action': it wants us to achieve, and will wake us up and focus our minds.

Vanadinite's dense appearance means it is an energy tool to help us navigate life. Its orange and red tones indicate a connection between the sacral and root chakras. When we look at our progression towards a goal, the sacral chakra is active in the creative stage and the root chakra represents the end result. It gives us the confidence to tell the world what we have achieved.

HOW TO USE IT

Display Vanadinite in the space where you will be working on your projects. If you are experiencing lots of limiting beliefs relating to your goal, first consider if it's the right goal for you. If it is, consider making an indirect crystal elixir (see page 46) with your Vanadinite.

Some people might find that Vanadinite is a little too forceful for them. If that is you, consider sourcing a specimen that has Barite as well – this softens the energy.

Additional properties: Goal setting, creativity and discipline

Sources: Morocco, Mexico and Arizona

Crystal shape: Raw

Best times to work with this crystal: Tuesday, afternoons

JULY GODDESS CRYSTAL:
TURQUOISE

GOOD LUCK, PROSPERITY AND PROTECTION

The goddess of precious metals and gems, Hathor protects miners and serves as patroness to the gem mines in the Egyptian Sinai peninsula.

She was believed to have loved gemstones, specifically Malachite and Turquoise, and was named lady or mistress of Turquoise, making it a highly significant crystal in ancient Egypt. You might remember Tutankhamun's iconic gold burial mask, inlaid with Turquoise, Lapis Lazuli, Carnelian and coloured glass.

Hathor, along with Turquoise, soon gained an association with good luck, prosperity and protection. Burning herbs and petals as an offering is an ancient way to ask a deity to bless our goals and ensure a good outcome.

HERBAL BLESSING TO HATHOR

When I host moon ceremonies, I often finish by burning some herbs to wish those who attended and their loved ones a joyful few weeks before our next gathering. We can also offer these blessings to the deity we are cultivating a relationship with, in return for their support.

You'll need a glass, one or a selection of herbs that symbolize good luck – such as thyme, sage, chamomile and rosemary – dry petals, a Turquoise crystal, a glass, a cauldron or other heatproof container, a charcoal tablet, tongs and matches.

1. The night before the blessing, put some of the herbs, petals and your crystal in the glass. This allows the crystal to endow its energy to everything else.

2. On the day, place the cauldron in a place where items can be burnt in it safely.

3. Remove your crystal from the glass. Hold the charcoal tablet with the tongs, light it and place it in the cauldron.

4. With the intention that you are making an offering to Hathor in return for good luck, sprinkle in the herbs and petals and watch them smoke.

Additional properties: Ancient wisdom, intuition and ancestral connection

Sources: The US, Iran, Mexico, Chile, China and the UK

Crystal shapes: Carvings, jewellery and raw

Best times to work with this crystal: Monday, lunchtimes

JULY CRYSTAL OF THE MONTH: CITRINE

PURE JOY AND HAPPINESS

I love this crystal – its energy is so positive and inspiring. Citrine is like the motivational speaker of the crystal world. Its predominant energy is 'pure joy', one of the highest vibrational frequencies we can strive to reach. To get you to this emotional state, Citrine sees its job as bringing true happiness into your life. This isn't just material possessions, but the knowledge, wisdom and skills that will help you achieve your life's purpose. That means Citrine is also an amazing manifester.

Citrine's ability to support our greater soul purpose, its sunny yellow tones and its positive, joyous nature, makes it the ideal representation of July's energies.

HOW TO USE IT

I recommend wearing Citrine as much as possible so you can enjoy its uplifting energy. Ideally, wear it on a necklace with a long chain so it is close to its associated chakra, the solar plexus. Once a week, hold it in your dominant hand as you visualize your dream life. This gives you and your Citrine a small reminder of what you are working towards.

Along with manifesting work, Citrine can help illuminate any blocks or the limiting beliefs that might stop you reaching your goals. For this kind of support and to motivate you to achieve your goals, take a Citrine elixir daily.

There are two types of Citrine in crystal shops. Natural Citrine is a rare form of Quartz, and this rarity is reflected in the price, which motivated people to create the second form of Citrine, burnt Amethyst Citrine. This is Amethyst that has been heated until it turns yellow. Both share the same properties, but some people will prefer natural Citrine, while others enjoy burnt Amethyst Citrine because of the creative aspect involved in its making.

Additional properties: Happiness, success and confidence

Sources: Brazil and Madagascar

Crystal shapes: Jewellery, points, spheres, wands and freeforms

Best times to work with this crystal: Tuesday or Sunday, afternoons

JULY ACTIVITIES

1. LION'S GATE ACTIVATION

One reason Leo season can feel especially intense is because from the 28 July to 12 August a series of astrological events occur. This includes an alignment between the sun and the star Sirius, giving us a closer connection between the physical and spiritual worlds. Orion's belt also aligns with the great pyramids of Giza, which, combined with the heat of the sun, intensifies all our energies. This peaks on 8 August (8/8), a number that symbolizes the infinity spiral of life, death and rebirth. All of these occurrences create a unique window of time known as the Lion's Gate, during which many people either realize their soul's purpose, or feel as though they are being moved towards it faster.

MERKABA VISUALIZATION

A merkaba is a Platonic solid, a sacred geometry shape that appears throughout nature and the universe. Some believe that our souls are brought to Earth inside a merkaba, and we can work with this shape when searching for guidance about our purpose. On 8 August, ideally at 8am or 8pm, do this visualization to help you understand what your soul's purpose is, or help you progress further down that path. You will need a Clear Quartz merkaba crystal, any size.

1. Sit, holding your crystal in your non-dominant hand, close your eyes and relax. Imagine a white light coming from the ground beneath you, travelling up into the body, up the spine, through your head and collecting at the crown chakra.

2. Now imagine a sphere of magenta light 1 metre (3 feet) above you. This is your soul star chakra, and is an aspect of your soul energy. Let the light collecting at your crown travel up to the soul star and stay there for a while. Have the intention to bring some of that light back down towards you, but this light now includes guidance on your soul's purpose. Let it enter your head, travel through your body to the hand holding your crystal, and then into the crystal itself.

3. Once you have programmed the crystal with your soul's purpose, display it somewhere you regularly sit so that you are around its energy and can pick up on any messages it might have.

2. RAISE YOUR VIBRATION

It is probably clear by now that the summer sun and the magical alignments in the stars this time of the year are energies we should harness. These heightened energies can also be used to amplify the energies of our crystals. You might know about cleansing and charging your crystals in the moonlight, but did you know that the sun can do the same?

On page 118 we explored Citrine and saw how its main property is to help people feel pure joy and happiness. This is the highest vibration a person can experience. This is the same vibration as enlightenment, which is a result of raising our vibration from what is known as third dimensional vibration, the physical world we live in, to fifth dimensional, a more compassionate and globally aware mindset.

Some people believe that this change in frequency is the next step in our evolution, the result being a happier, heart-expanded international community who create a more community focused, altruistic world.

RAISE YOUR VIBRATION WITH CITRINE

1. Start by leaving your Citrine crystal out on the windowsill for a few days so it can be charged by the sunlight.
2. On a sunny day, apply sunscreen and head outside to find a spot where you can lie in the sun without being disturbed.
3. Once you're lying down, place your Citrine crystal on your solar plexus, then place your hands out to the side (so your body creates a star). Imagine a ball of yellow light coming out from your crystal and surrounding you.
4. Stay there for 5–10 minutes, then allow the ball of light to go back into the crystal.

August

SUMMER ENDS: **STAMINA, ACHIEVEMENT & EVOLUTION**

In this, the final month of summer, we aim to get outside as much as possible to enjoy the remaining warm days while we still can. August also marks the final push to finish our biggest projects.

Lughnasadh (pronounced *loo-nah-sah*), also called Lammas, is celebrated on 1 August. This sabbath marks the first harvest: after months of tending to the fields, it is a day of celebration before the back-breaking work begins. It is a community event – everyone is working together to gather the crops before the weather changes and they are lost. Today we are more likely to be found working on our big goals from a home office. After spending months on one project, the finish line is in sight, but momentum might be waning.

In late August, when we move into Virgo season, our energies go from doing and achieving to completing. At the same time, we see the first signs of autumn. The countdown to cold days and longer nights has begun. We will need to spend more time indoors soon, but not quite yet – there's still time to play.

This month we use crystals to honour the Greek deity Gaia, to strengthen our connection to nature. I also suggest crystals to reinvigorate your goals and help you conduct some inner-child healing to ward off negative thinking.

August facts
Colours: Red, orange and yellow
Crystal shapes: Cubes, flames, merkabas and sacred geometry sets
Seasonal chakra energies: All seven chakras
Animals: Tigers, ants and starfish
Seasonal shrine/altar: Music, citrus fruits, corn dolls and bread
Intention setting/affirmation: *The end is in sight*
August birthstone (traditional): Sardonyx
Birthstone properties: Determination
August birthstone (modern): Peridot
Birthstone properties: Positivity

AUGUST SABBATH CELEBRATION:
LUGHNASADH OR LAMMAS

There are three harvest festivals on the wheel of the year, and the Irish festival of Lughnasasdh is the first. This festival was named after the Celtic sun god, Lugh. Little is known about this god, as Irish mythology was rarely written down, but when the ancient Romans came to the area they likened Lugh to their god Mercury, so Lugh was then considered a warrior and was portrayed as a god of craftsmanship.

Irish folklore suggests the celebration of Lughnasadh either marks his marriage or honours Lugh's foster mother, Tailtiu, who died of exhaustion when clearing land in Ireland for agriculture. Either way, the festival gives thanks to their sun god and would mark the date the first grains can be cut: crops couldn't be harvested before this date because they wouldn't last the winter. Work started at dawn, and by dinner the first loafs were baked and enjoyed with a well-deserved glass of wine.

The festival Lammas is more closely connected to the time the harvest festival melded with Christianity in the early Middle Ages. In this case, the first loaves of bread would be taken to the church to be blessed and offered at mass. Lammas is an Old English word for 'loaf mass'.

For many people, the traditions have merged and are celebrated together on 1 August as one of the four cross-quarter celebrations known as fire festivals. This one marks the halfway point between the summer solstice in June and the autumn equinox in September. Each festival also honours the element connected to that season, which this time is Fire – another example of how the Fire element burns particularly bright throughout the summer months.

CRYSTALS TO DISPLAY FOR LAMMAS

Enhancing the sun's energies: Sunstone and Fire Quartz

Creating willpower: Tiger's Eye and Vandanite

Offering gratitude: Mangano Calcite

AUGUST ASTROLOGY
& FULL MOON

We progress into August with Leo still asking us to go full-steam ahead: work hard and play hard are the messages this star sign is still sending us. However, around this time we remember the lessons of the spring equinox, urging us to stay balanced through the summer so we don't burn out. You might find this time of year intense, but at the same time Leo season is perfectly positioned to offer us a final injection of motivation to get our goals over the finish line.

On 23 August we move into Virgo season. Suddenly, the inner child in all of us wants to head indoors for a mid-afternoon nap. The projects we have been working on finally end and we display the fruits of our labour to the world. Virgo is an Earth element sign and Virgo energy tends to jump ahead to evaluate the success of a project too soon, so make sure you take time to celebrate what you have created.

Virgo's planet is Mercury, the planet related to communication, so it shares its insights through clareaudio (intuitive thoughts). The problem is, after a busy few months, we might be tired, and any insights that feel like criticism could demotivate and stop us progressing. When this happens, remember: it's not your inner critic speaking, just misguided but caring Virgo offering suggestions because it wants us to make well thought out, grounded decisions that will last.

For this reason, Virgo season is the perfect time to plan an away day so you can consider what went well and how things can be done differently in the future. That way, you are welcoming in Virgo's opinion for a focused amount of time, so it doesn't need to offer its reflections all month.

AUGUST'S FULL MOON

This month's full moon is often called the Sturgeon Moon because this fish was a major food source for some Indigenous Americans around this time. The alternative name of Corn Moon refers to the seasonal celebrations taking place.

AUGUST GODDESS: GAIA

NAME AND MEANING

Gaia (also called Ge, Mother Earth or Terra) is the first deity in Greek mythology, the creator of the Titans and of all life. Her name means 'Earth'. She is honoured all year round, but particularly in the height of summer when we are outside enjoying everything she has created.

TEACHING

Greek myths explain that Gaia came first, then her counterpart Ouranos (Father Sky). Together they play out the creation myth: Gaia birthed the land and Ouranos sent rain to form the oceans, from which life emerged. Then they separated and Ouranos formed the skies. One by one everything was formed, even the monsters from the famous Greek legends. It was believed everything that exists owes its origin to the mother goddess, Gaia.

CHARACTERISTICS

Gaia is depicted in the famous Venus of Willendorf, the oldest known sculpture of the human form. Later she would often be represented rising from the ground, surrounded by fruits, vegetables and flowers; the ancient Greeks gave Gaia offerings of honey cake in return for a good harvest.

It is no surprise that ancient and modern cultures all around the world have their own Earth goddess, from the Mayan fertility goddess Pachamama, to Houtu the Chinese queen of the Earth, Wicca's triple goddess and modern images of Mother Nature depicting the Earth as her pregnant belly.

Terra is believed to be Gaia's Roman counterpart. Along with the goddess Ceres, the goddess of agriculture and grain crops, they both would have been celebrated and honoured at this time of the year.

SYMBOLS OF GAIA

The feminine form, the spiral goddess, honey cakes and planet Earth

AUGUST SABBATH CRYSTAL:
FIRE QUARTZ

PURPOSE

We are in the Fire element season: the sun is high in the sky and has been for several months, building energies. When August begins we are still in the Fire sign season of Leo and are enjoying the energies of Lion's Gate. Lughnasadh celebrations happen when those energies are at their peak.

A crystal that compliments these heightened energies is Fire Quartz, which contains Red Hematite found in the Quartz and amplifies the Fire element and our life passions.

No matter what project you have been working on this year, by now your energies might be waning and you might feel the need to slow down. To get you across the finish line you can maximize the energies of the Fire element to keep your stamina up.

HOW TO USE IT

Here is a simple crystal layout that is perfect for Lammas, and will strengthen your auric field (the part of your energy body that surrounds your physical body). You will need enough Fire Quartz chips to create a circle large enough to lie inside. Relax inside the circle for up to 15 minutes while the crystals work on strengthening the energies around your body.

Additional properties: Passion for life, motivation and drive

Sources: The USA, Brazil and Madagascar

Crystal shapes: Chips, tumbles, towers and flames

Best times to work with this crystal: Tuesday, afternoons

AUGUST ASTROLOGICAL CRYSTAL:
INNER CHILD QUARTZ FORMATION

INNER CHILD HEALING

The different shapes formed by Quartz crystal also have their own properties and specific meanings. A Quartz crystal that has another crystal growing inside it with one end sticking out is called an Inner Child, or Penetration Quartz. The central point on top of the crystal in the photograph opposite is an inner child.

True to its name, this crystal represents aspects of us that have stayed frozen at a certain age, normally under seven years old. This can happen when one or multiple traumas or life events have happened at a young age.

As an adult, that part of us might react negatively to situations in a way that is out of character and could be childlike. The inner child is essentially the voice of unhealed emotional wounds from childhood and, as Virgo season can create an increase in negative internal dialogue, it can help us see what needs healing.

Even if you don't experience these voices at other times of the year, you might hear them now. It can feel like tough love, but it happens once our goals have been achieved to help us evaluate how everything went. We then have the autumn and winter seasons to heal or make change, ensuring that next year we can build on our current success.

HOW TO USE IT

Start by displaying your inner child crystal throughout the Virgo season. This projects healing energies all around you, working subtly all the time. If you would like to dive deeper and progress your inner child healing further, see page 135.

Additional properties: Trauma release, ancestral healing and healing co-dependency

Source: Worldwide

Crystal shape: Inner child quartz formation

Best times to work with this crystal: Friday, mornings

AUGUST GODDESS CRYSTAL: GARDEN QUARTZ

CONNECTION TO NATURE

A form of Phantom Quartz, Garden Quartz stands out because of the formations of the minerals (inclusions) inside the crystal. Also called Lodalite, Landscape Quartz or Inclusion Quartz, Garden Quartz contains a vast variety of crystals and minerals, including Chlorite, Feldspar, Iron, Magnesium and Mica. Together, the inclusions form the appearance of miniature worlds; some look like gardens, others like rolling hills, savannahs or forests.

Normally I would suggest a dense crystal to represent the natural world, but the contrast of transparent etheric Quartz next to the dense inclusions is the perfect representation of an Earth goddess crystal. The natural formations inside means it's a crystal that can connect us to nature. When we feel this connection, we naturally feel calmer and more grounded.

Garden Quartz specimens normally come in Smoky or Clear Quartz, but it is possible to find natural Citrine in Garden Quartz as well. Each type of mineral adds its individual property to the general properties of Garden Quartz.

Clear Quartz: Clarity

Smoky Quartz: Guidance

Natural Citrine: Joy and abundance

HONOURING GAIA

If you are attracted to crystals, you are already actively cultivating a relationship with Gaia. To give thanks to her for your beautiful crystal collection, I suggest sourcing a Garden Quartz crystal you are particularly drawn to, which can take pride of place in your home. Then you might like to display a collection of your favourite crystals around it. Consider displaying different shapes and colours to showcase a spectrum of Gaia's creations.

Additional properties: Grounding, balancing and manifesting

Source: Brazil

Crystal shapes: Towers or spheres

Best times to work with this crystal: Friday or Saturday, mornings

AUGUST CRYSTAL OF THE MONTH: CARNELIAN

THE CREATIVE PROCESS

Carnelian can come in one colour, or in a mix of red, orange and yellow colours. Its connection to the full spectrum of fire colours embraces the energy of the summer months, but with less intensity than Fire Quartz or Fire Opal. Carnelian contains the colour of the three lower chakras. Together they represent the creative process, the actions we need to undertake to achieve a goal:

Yellow: Use on the first day of a project when vitality and energy are needed to begin. Select a predominantly yellow Carnelian if you have a tendency to procrastinate.

Orange: The creative, 'doing' stage. If this is a long project, choose a crystal containing lots of orange to give you the willpower you need to reach the finishing line.

Red: This represents the outcome. If you find it hard to talk about your achievements, choose a crystal with more red in it as this will help to ease that process.

HOW TO USE IT

Source a Carnelian specimen that matches your needs. I recommend a large tumble you can keep in your pocket, or a ring. That way it will sit close to the relevant chakras. Keep that crystal with you throughout the summer months, but especially in August when the energies are heightened.

Additional properties: Productivity, creativity and fertility

Sources: India, Brazil and Uruguay

Crystal shapes: Tumbles, points or stars

Best times to work with this crystal: Tuesday or Thursday, afternoons

AUGUST ACTIVITIES

1. CRYSTAL CANDLE

When summer ends, our goals are also reaching their end point, and you may wish to celebrate. Here I show you how to make a celebratory crystal candle.

Creating candles decorated with crystals helps you connect with the Fire element, and will illuminate the energy of the crystals. Here I suggest making a Citrine candle for a joyous, celebratory theme.

MAKE A CITRINE CANDLE

You will need 185g (6½ oz) beeswax, a TCR wick, a 300ml (½ pint) glass candle jar, an old saucepan, a hot plate or stove, a thermometer, a metal candle-pouring jug, essential oils of your choice (optional; lemon oil is my recommendation), a wooden spoon, one or more Citrine tumbles, Citrine chips (optional) and dried daisies or sunflower petals (optional).

1. Heat some water in the pan until it reaches 60–70°C (140–158°F). Place the wax in the jug and carefully put the jug in the pan. Let the wax slowly melt, stirring as it does. Once the wax has melted, turn down the heat. Using a few drops of wax, fix the wick to the bottom of the jar.

2. When the wax temperature is at 55°C (131°F), add up to 18g (²/₃ oz) of essential oils (optional) and stir in. Using an oven glove, take the jug out of the pan and let it cool. When it is between 45–50°C (113–122°F) you can fill roughly 80 per cent of the candle jar with your wax. Tap the jar a few times to release air bubbles.

3. Let it cool enough so that the wax can support the weight of the tumbles. Carefully add the crystals and petals, then pour in the rest of the wax so it surrounds but does not cover the crystals. Trim the wick.

> You can also light a candle you already have, let it burn for a while and then add crystals to the pool of melted wax.

2. INNER CHILD CRYSTAL HEALING

We have seen (see page 128) how your inner critic's activity can increase during Virgo season because Virgo shows us what still needs to be healed. Any healing connected to our inner child will make a big positive change in our life, and you can facilitate this healing with an inner child crystal formation.

HOW TO DO IT

When you want to focus on a specific emotion or behaviour, start by journalling on that topic. Consider writing about how these behaviours play out in your life and what the emotions feel like and when they arrive. Also consider how it makes you feel, what you wish would happen instead and when you think those emotions/behaviours first started.

1. While you journal, see if the emotions feel like they're coming from a specific area of the body, or observe whether the body heats up, starts to tickle, if your throat becomes scratchy or there is some form of mystery pain. If it does, place your inner child crystal over that space, close your eyes and relax. Take a few breaths.

2. On an inhale, ask your crystal what emotion is there, and let it give the name on the exhale. Following the breath again, ask the crystal why that emotion is there and if it has a message for you. If you don't get an answer, try a few times and then move on, as this step is useful but not essential.

3. Now turn that emotion into a colour and see that colour move through the body into the crystal. Move that crystal away from the body and cleanse it of the old energies.

Cleansing Your Inner Child Crystal: For an Earth element star sign like Virgo, I would suggest using an Earth element process for the cleansing. The most profound way is to bury your crystal in the ground for a few days, then dig it up again. If that isn't possible, you can cover it in rice for a few days, but make sure you don't consume the rice afterwards.

September

AUTUMN BEGINS: **INSIGHTS, CALM & REBALANCING**

As the clock strikes midnight on 1 September, meteorological autumn arrives. We go from days and night outside to observing the first leaves falling from the trees. There is a collective sigh of relief. The busy summer months are over. We are allowed to relax, to go back to our regular work routine and allow ourselves time inside to rest.

Autumn is connected to the Water element and our emotional body. Virgo season is still asking us to consider how our goals went and what needs changing or healing before next year's goals begin, then Libra takes control and reminds us how important balance is.

This month we celebrate Mabon, the second harvest festival of the year and the autumn solstice, when night and day are momentarily equal in length for the second time. The energy of balance is felt globally again. Moments later, the northern hemisphere moves into autumn and the southern hemisphere welcomes spring. Fruits and seeds are everywhere; once they're picked or fall to the ground the tree's job is done and its slow retreat can begin.

This month we use crystals to connect to the balancing yin and yang energies of this time with Chinese goddess Xiwangmu, introducing more harmony to our personalities, body and relationships. I also show you how to use crystals for manifesting your goals and introducing more structure into your routine.

September facts
Colours: Brown and orange
Crystal shapes: Jewellery, statement pieces, pyramids
Seasonal chakra energy: Heart chakra
Animals: Squirrels, zebras and horses
Seasonal shrine/altar: Cornucopia, acorns and orange candles
Intention setting/affirmation: *I am balanced in all areas of my life*
September birthstone (traditional and modern): Sapphire
Birthstone properties: Wisdom and higher forms of communication

SEPTEMBER SABBATH CELEBRATIONS: AUTUMN EQUINOX & MABON

The autumn equinox is the second equinox of the year, and it normally occurs on 21 September, though it can vary by a day either side. This is also the date of the Mabon celebrations. Like the spring equinox in March, this marks the moment in time when night and day are equal.

In late September, farmers are still working hard, harvesting their crops. Mabon is a moment of celebration for them and their communities as the tree, berry and vine fruit harvest comes to an end. At the same time, supplies are being prepared and preserved for the cold months ahead.

It was an intense moment for our ancestors, who would know by now how well the year's harvest had gone. Like an echo of the past, we can only imagine the emotions they would have felt when they realized if the winter was going to be easy or a struggle. Autumn is connected to the Water element and our emotions, and these moments are a perfect example of this.

No matter what, it was still important to celebrate and give thanks to the gods, offering gratitude for a plentiful crop or placating them if stocks were low. The writer on Wicca Aidan Kelly officially renamed this festival Mabon after Mabon ap Modron, a character from Welsh mythology who was closely connected to King Arthur. Still an important celebration for pagans and Wiccans, it is often called the witches' thanksgiving.

As this is the second harvest celebration, there is a feeling of abundance in the air, and it's important to give thanks to Mother Earth for her support, and to share the harvest with all the community so those energies of abundance are extended and will attract the same luck the following year.

CRYSTALS TO DISPLAY FOR MABON

Create balance: All Agate crystals

Offering gratitude: Mangano Calcite

Good luck: Turquoise and Green Jade

SEPTEMBER ASTROLOGY
& FULL MOON

 Virgo brings us into September, which makes perfect sense; many people are going back to school or settling back into work after summer breaks, so looking at our schedules and planning ahead is everyone's focus.

 Our ancestors would have taken advantage of the Virgo energies to plan and implement the harvest. Libra then steps in around the equinox on 23 September. Its symbol is the scales, enhancing the equinox's natural ability to create balance, allowing that energy to extend several weeks this time.

Even though our daily activities have changed, these energies are still the dominant focus for most people, just as they would have been hundreds of years ago. Nowadays, balance might mean changing our diet or adding a self-care routine after the busy summer months, but historically this would have been preparing stocks and planning rations for the coming winter.

Libra is an Air sign, which sets the perfect stage for planning and creating a spreadsheet or two. It's also a cardinal sign, so it can take charge and make rules. It's highly diplomatic, and able to find a way to help everyone understand when rations were particularly tight. Libra's planet is Venus, the planet of the physical world and material possessions, so if you feel moved to decorate or change the home around during the Libra session, that could be why.

SEPTEMBER'S FULL MOON

Normally the name of a full moon will be the same no matter when it occurs in the month, but not this time. The full moon that is closest to the equinox (so it could be in September or October) is called the Harvest Moon. This moon rises around sunset each year for several days in a row, allowing people who are working on the harvest to get a few extra hours of work done. When September's full moon isn't a Harvest Moon, it is called the Corn Moon, a name sometimes given to August's moon in years when September has the official Harvest Moon.

SEPTEMBER GODDESS: XIWANGMU

NAME AND MEANING

Xiwangmu's story evolved from that of a vengeful goddess of death depicted with big cat teeth who sent wild animals and diseases to destroy mankind – but she became the highest-ranking female deity in the Chinese pantheon. Her name means Queen Mother of the West, and she is also the goddess of good fortune, long life, immortality, yin (feminine) energy and female empowerment.

She was originally said to be married to Dongwanggong, the personification of masculine yang energy. Later, as her story evolved, Xiwangmu was the consort to the Jade Emperor, ruler of the heavens, with whom she had many children.

CHARACTERISTICS

Xiwangmu lives in a beautiful celestial place at the top of Mount Kunlun. From there she watches over all other female deities, guiding and teaching them. She helps women reach enlightenment, guiding them along their spiritual path. Perhaps that is why she is often depicted wearing a peach headdress. Grown in her magical garden, these peaches ripen once every 3,000 years. Eating one offers you immortality, so Xiwangmu restricts access to a chosen few.

TEACHING

Xiwangmu had a particular fondness for women who wanted to experience life outside the cultural norms of the time, and they loved her back equally. Still a popular deity today, particularly among Daoists, Xiwangmu champions the importance of divine feminine, understanding that women are men's equal, not subservient. This is why Xiwangmu is our September goddess. As the Earth falls back into equilibrium, she reminds us that balance is needed in all areas of life, including equality. This process begins by honouring and balancing the divine masculine and feminine in each of us.

SYMBOLS OF XIWANGMU

Peaches and the Big Dipper

SEPTEMBER SABBATH CRYSTAL: RUTILATED QUARTZ

MANIFESTING

After a bountiful harvest, Mabon is the perfect time to give thanks. But even after a difficult harvest, our ancestors were grateful in order to placate the gods and goddess so they would receive a better harvest the following year.

In reality, actively being grateful for whatever is in your life raises your vibration so that more good things can enter. Rutilated Quartz (also called Angel Hair Quartz or Venus' Hair Stone) is an expert manifester because it has already manifested/created the rutiles, or needle-like formations, inside the Quartz crystal. It has a busy, active energy, perfect to help motivate you to achieve new things.

Rutilated Quartz can come in the form of Clear Quartz, Smoky Quartz and natural Citrine. The rutiles inside also come in multiple colours, which have different properties.

Red and brown: These reduce negative thoughts by raising our frequency above them so they cannot hinder our progression.

Orange and gold: These support creativity and put an emphasis on goals that are connected to our soul's purpose.

Silver: Enhances our intuition to help us receive guidance.

HOW TO USE IT

Treat yourself to a Rutilated Quartz pendant that you're drawn to, which also has coloured rutiles that suit your needs. Wear it on a long chain so it's close to your solar plexus; this will help you focus on your goal.

Once a week, lie down and place the crystal between your heart and solar plexus. Visualize your goal and imagine you have already achieved it. Ask yourself how that feels. Then imagine a swirl of energy coming from the crystal, spiralling out, connecting to the image of your goal and bringing it to you.

Additional properties: Focus, goal setting and drive

Sources: Brazil and Madagascar

Crystal shapes: Jewellery, points and tumbles

Best times to work with this crystal: Tuesday and Thursday, afternoons

SEPTEMBER ASTROLOGICAL CRYSTAL: AMAZONITE

DIPLOMACY

Libra asks for balance in all areas of our life, but with its connection to the planet Venus, balance in relationships is specifically important to this sign. This makes September the perfect month to evaluate your relationships and consider which ones might be out of balance. Where are you giving more of yourself than you're receiving? Or are others supporting you more than you are supporting them?

This is the perfect time to carry Rose Quartz with you, as it will heighten the focus on your relationships and help you understand where work is needed.

Sometimes conversations can bring relationships back into balance, but this can be hard, especially if that relationship is complicated. If this is the case, we can work with Amazonite. This is also called Amazon Stone or Amazon Jade, after the famous Amazon warriors, who were skilled fighters – not that they needed to fight, as their diplomacy skills were second to none.

Amazonite combines blue, reflecting their strong communication skills, with green, enforcing the energies of the heart chakra. Together they can help us speak from the heart, allowing us to communicate to someone in a way that won't cause them offence.

HOW TO USE IT

Between the heart and throat chakra is the lesser-known higher heart chakra, which offers a more expansive and compassionate heart energy. It is this energy that Amazonite taps into.

When entering into a difficult conversation, wear an Amazonite necklace positioned over the higher heart. This will help you access this chakra so you can communicate from that space.

Additional properties: Courage, understanding and fairness

Sources: Brazil, the USA, India, Madagascar, Namibia and Russia

Crystal shapes: Jewellery, hearts and palms

Best times to work with this crystal: Saturday, afternoons

SEPTEMBER GODDESS CRYSTAL: MOQUI MARBLES

BALANCING YIN AND YANG ENERGIES

This way of working with crystals combines two ancient cultures: ancient China and Indigenous American culture, specifically the Hopi tribe, originally called the Moqui. You will remember that goddess Xiwangmu teaches that all genders should be seen as equal, and she was the embodiment of yin energies. With the arrival of the autumn equinox and Libra season, we are asked to see where in our life equilibrium is needed. Balancing the masculine and feminine aspects of our energy and personality is a part of that, and Moqui Marbles are the perfect crystal tool to help us achieve it. They are also known as Shaman Stones, Navajo Cherries, Hopi Marbles and Moqui Balls.

Found only in the Navajo region of Utah and Arizona, Moqui Marbles are a concretion of sandstone balls cemented by a hard shell of iron oxide minerals. These are ancient stones, found freely in the Navajo desert. A sacred stone to the indigenous peoples, they were used in spiritual practices, as well as games.

Interestingly, they form in two shapes: disc-like formations that give off energies of the divine masculine, and spheres that offer the energies of the divine feminine. Therefore it is always a good idea to source a set to work with. If the shapes are uneven, you can use a pendulum to decide if you have a masculine or feminine crystal.

HOW TO USE THEM

Here is a simple practice to balance the male and female energies in your body.

1. Sitting in meditation, place the male crystal in your right hand and the female crystal in your left hand. Stay there for 5–10 minutes. You might like to keep your focus on the breath, inhaling and exhaling to the count of four or five, holding the breath at the top before you exhale.

2. When you are done, place both crystals by your feet. Moqui Marbles are also good grounding crystals. Leave them there for a few moments so they can ground your energies before you stand up.

Additional properties: Duality, unity and honouring both ying and the yang

Sources: Arizona and the USA

Crystal shape: Raw

Best times to work with this crystal: Saturday, mornings or evenings

SEPTEMBER CRYSTAL OF THE MONTH: BISMUTH

ORDER AND STRUCTURE

September is about transition: going from an outdoor lifestyle to indoors, holiday season to work and school, from hot days to colder weather.

Libra and the autumn equinox ask us to make sure all areas of our life are as balanced as possible, and reminds us that planning is needed if we want this transition to be easy.

Bismuth starts as a crystallized white metal that, through manmade intervention, becomes a crystal. Starting life as the chemical element Bi on the periodic table, this metal was one of the first to be discovered. It melts at a low temperature, and once molten it reorganizes its molecules into interesting shapes, similar to step-like pyramid structures. This is known as a Hopper crystal shape. At the same time the oxidation that takes place creates a rainbow aura effect.

To officially be classified as a crystal, a structure's molecules need to appear as a regular repeating pattern, so Bismuth transforms from a metal into a crystal, meaning it also goes from the unorganized atomic structure of metal to an organized and visually striking crystalline structure. Its metamorphosis offers us the inspiration and motivation to restructure ourselves. The rainbow effect is created by oxidation, indicating how Bismuth can support all our chakras and therefore all areas of our life.

HOW TO USE IT

Display this crystal in an area of the home or workspace where most of your work takes place. That way you can pick up on its transformative energies as you work so the process feels easier and flows better.

Additional properties: Reconstruction, bravery and aura cleansing

Sources: South America and Canada

Crystal shape: Raw

Best times to work with this crystal: Saturday, all day

SEPTEMBER ACTIVITIES

1. BALANCING THE ENERGY BODY WITH CRYSTALS

When you use any energy healing modality, you are working with the energy body. Your chakras are part of this, as is your entire auric field, the sphere of energy that surrounds us all. It is easier to keep our energy body healthy than our physical body because there is greater space between the atoms to facilitate change and healing. What's healed on that level then feeds into our physical body, aiding the healing of our physical body as well.

The theme of this month's energy is balance, and our energy bodies like to stay in balance as well. On page 144 we saw how Moqui Marbles balance masculine and feminine energies; we saw how to balance your chakras on pages 20–1. Now I'm going to show you how you can use crystals to balance the elements in the body.

BALANCING THE ELEMENTS
We are made up of all the elements. Over time, we can become energetically depleted in one or more of them. To attract more of that element, use a crystal point (raw or polished) that is connected to the element. Try Clear Quartz for Air, Aquamarine for Water, Hematite for Earth and Fire Quartz for Fire.

1. Create a circle you can lie down inside, with the four crystals around you, points facing in (so they pull the element towards you). Position the crystals roughly in the correlating location: north for Air, east for Water, south for Earth, west for Fire. Note that these locations correlate to the teachings I have received; different cultures connect the points to different elements, so choose what resonates with you most.

2. Lie in the centre of the circle with the Earth crystal below your feet and the Air crystal above your head. Relax there for 20 minutes. At that point take away the Air, Water and Fire crystals, leaving just Earth in place. Stay there for another few minutes to ground your energies.

2. CRYSTAL INTUITION

Some people may hate the idea that September is a month for planning and preparing, preferring to make spontaneous decisions and live in the moment. Do you prefer using your intuition to guide you? If this is you, I would suggest taking a different approach to Libra's analytical planning. Instead, you might see September as a time to welcome in intuitive insights that specifically offer up guidance on action you need to take in the next few months.

When we sleep, our conscious mind is calm, allowing the subconscious to come through better. If you wake up in the morning with ideas and inspirations you want to pursue, this is why.

USING CELESTITE CRYSTALS FOR GUIDANCE

Throughout September, consider displaying a Celestite crystal by your bedside. This crystal aids sleep problems, so it will naturally facilitate deep sleep. As its name suggests, it is also one of the crystals we can use to help us connect to our guardian angels. Displaying your crystals by your bedside table welcomes their guidance. Each night before you go to sleep, look at your crystal and invite it to offer you wisdom. Over the month, in your waking state you will find plans are made and ideas come together, thanks to the help of this crystal.

October

MID-AUTUMN: **ANCESTORS, PROTECTION & INTROSPECTION**

Autumn is here and nature is dying back. Trees decorate the forest in red, orange and yellow. Carved pumpkins, Halloween decorations and horror movies on TV enhance October's ghostly energies. The weather has also shifted. Fog and mists linger in the empty fields, the result of the colder nights. Students and workers head home once their working day is over, prioritizing home comforts over socializing.

Libra season welcomes October, asking for more balance in life. As its energies build, we recognize which of our relationships are out of balance. From the middle of the month, these lessons lead us into Scorpio season. Along with the increasingly dark nights, this is when the months governed by the Wise Woman goddess begin and we start our annual journey looking inwards: healing.

The celebrations of Samhain and Halloween marked a time of genuine fear for our ancestors. For them, this time of year had a sinister aspect, as Samhain marked the doorway into the uncertainty of winter. This is a moment to honour our ancestors: because they found a way through, we are here today.

I show you how to channel this month's magical energies by using crystals to connect to intuitive messages from beyond with the ancient Greek goddess Hecate, protect your energy and connect with your emotions.

October facts
Colours: Black, brown, orange, purple and silver
Crystal shapes: Tumbles, raw, wands, hearts and skulls
Seasonal chakra energies: Root and crown chakras
Animals: Cats, bats and scorpions
Seasonal shrine/altar: Jack-o'-Lanterns, family photos and orange candles
Intention setting/affirmation: *I am grateful to those who came before me*
October birthstone (traditional): Opal
Birthstone properties: Emotional support and purity
October birthstone (modern): Pink Tourmaline
Birthstone properties: Compassion

OCTOBER SABBATH CELEBRATION: SAMHAIN

Since the start of September, nature has been retreating, dying off and returning to the ground. By Samhain (pronounced sow-wen), the results of this are clear, and you might even feel the energy of death in the air.

Samhain celebrations start on the eve of 31 October and carry on into 1 November. Originally a Celtic celebration, Samhain is an old Irish word for 'summer's end', the final day of the harvest. It was unlucky to take anything after this date, instead those items needed to be gifted to the fairies.

This festival marked the movement when livestock were moved to their winter pastures. Some would have been killed, their meat preserved, adding another layer of death to this festival.

This celebration was seen as the final hoorah before rationing really began. Any produce that wouldn't last the winter would be enjoyed.

Like Beltane in May, the veil between the living and the dead is believed to be the thinnest on this night. Irish mythology talks of burial mounds being opened to enhance that connection, and those celebrating Mexico's Day of the Dead might take a picnic to the graveyard to eat with deceased family members.

Houses were cleaned to cleanse them, while candles and lanterns were placed in windows to guide the spirits and protect the home.

Divination was a key aspect of the day's celebrations and there were multiple ways to seek intuitive guidance. Some are well known, like tarot cards and pendulums; others are more unique, like young women baking dumb cakes, eaten in silence before bed, so they would dream about their future husband.

CRYSTALS TO DISPLAY FOR SAMHAIN

Protection: Black Obsidian

Developing your psychic skills: Amethyst

Connecting to our ancestors: Petrified Wood and Bloodstone

OCTOBER ASTROLOGY
& FULL MOON

Libra takes us into October, and it is still asking us to observe where our relationships are out of balance and what we can do to correct them. This is our astrological point of focus until it's time for Libra to hand over the reins to Scorpio on 23 October.

With what needs changing still clear in our minds, Scorpio leads us into a month of introspection and healing, so deep shifts can be made. Scorpio is a Water element sign, so emotions are going to be intense, passionate and confronting. The intensity is related to Scorpio's connection to Mars, the planet of war. Pluto is also connected to Scorpio, which reflects its relationship with the death and rebirth process. It is fascinated with transformation, and is often connected with the symbol of the snake shedding its skin.

This season will ask you to look deep inside, under the protective shell of the scorpion, and find any hidden emotions. Topics that can come up a lot this month are obligation, shame, judgement and regret. You will know which one needs work when you consider how those themes have played out in your life. Look for one that gives you a pang of pain like a scorpion sting. That is where you need to focus.

OCTOBER'S FULL MOON

In a year when October's full moon isn't the closest to the equinox (in which case it would be called the Harvest Moon), it is given the name Hunter's Moon. This is because the crops in the fields have been cut, so it's easier for hunters to find their prey.

OCTOBER GODDESS: HECATE

NAME AND MEANING

An important Greek goddess and the representation of the Wise Woman aspect of the triple goddess, Hecate is the perfect support to guide us through Scorpio season and Samhain. She is the goddess of magic, witchcraft, the night, the moon, protection, ghosts and necromancy, and her name means 'far reaching'.

CHARACTERISTICS

Hecate fears nothing, including the afterlife: she escorted Demeter there when she first searched for Persephone. Then she regularly joined Persephone on her annual journey into the underworld. She possesses the ability to work with and control ghosts, spirits and dark magic. Hecate is comfortable moving through the dark and often chooses to work at night, particularly on a dark moon when deep healing can take place and intuitive messages are easier to obtain.

Hecate is regularly depicted with a two-headed dog who would bark when a threat approached, and this makes her a protection goddess as well.

TEACHING

She is the goddess and guide to all witches and those who have chosen a magical path. To these people she offers her wisdom, normally during sleep. Associated with the image of a crossroads, Hecate helps us decide on a path when we don't know which direction to take. That could be why she was also depicted as a triple-bodied goddess; she is able to face multiple directions and foresee all possible options. She is often shown holding two torches, lighting the way ahead.

SYMBOLS OF HECATE

Torch, key, snake, dogs, crossroads and the Hecate wheel

OCTOBER SABBATH CRYSTAL: BLACK OBSIDIAN

DIVINATION

With so many possible ways to carry out divination at Halloween, I wanted to share one of my favourites. Scrying is the act of looking into or at something to find symbols and insights that answer any questions you might have. That includes tea leaves, watching a flame, looking for shapes in the clouds or casting runes.

BLACK OBSIDIAN SCRYING MIRRORS

Black Obsidian mirrors are normally circular or oval sheets of Black Obsidian. In essence it is a black glass mirror, and the aim is to look for shapes in the shadows it creates.

Sitting comfortably, ideally in a position that allows you to hold your mirror in front of your face (or use a stand), close your eyes and think of a question you would like to be answered.

Wait until you feel completely relaxed, then open your eyes and look into the mirror. Explore the mirror's surface slowly until shapes start to appear. At that point, ask yourself what that shape means to you.

Remember, sometimes you might not get an answer. The position of the mirror can make it hard to see anything, or you might only see your silhouette. If this happens, understand that this is part of the message, and ask yourself what it means in relation to your question.

Additional properties: Protection, shadow work and understanding the darker aspects of life

Source: Worldwide, in areas that have seen volcanic activity

Crystal shapes: Shapes and mirrors

Best times to work with this crystal: Saturday, evenings

OCTOBER ASTROLOGICAL CRYSTAL: AQUAMARINE

PURIFICATION

This month you might see tempers boil over as we try to understand which deep-seated emotions would like to shift and release. Scorpio season can see emotions come through like a fast-moving rainstorm. The key is to allow these emotions to come, sit with them for a while and allow them to be heard. When we do this, they often release surprisingly quickly.

You can work with Water element crystals this month to help ease the level and frequency of emotions you might be experiencing (see page 165 to read about other Water element crystals). Here, we are going to look at Aquamarine. As its name suggests, Aquamarine has a strong affinity with many aspects of the Water element, from an energetic connection to water itself, to sea mammals like whales and dolphins and stories of mermaids, Poseidon and Atlantis, to the belief that Aquamarine offers safe travel by sea.

HOW TO USE IT

Carry an Aquamarine tumble in your pocket so you can hold it any time emotions come to the surface. Imagine water falling over you, flushing away those feelings. It is also a good idea to keep your crystal nearby while you sleep, so it can keep working overnight.

Additional properties: Cleansing, communication skills and safe travel

Sources: Pakistan, Brazil, Africa and Russia

Crystal shapes: Raw, tumbles and jewellery

Best times to work with this crystal: Saturday, mornings

OCTOBER GODDESS CRYSTAL: LABRADORITE

MAGIC

From a distance, the dark-textured tones of Labradorite (also called Spectrolite) make it look like a normal stone, nothing special – but when we get closer and examine it, there it is! Like magic, Labradorite's flashes of colour move over the iridescent surface of the crystal.

This is Labradorite's magic, and it wants to use its magical abilities to help you find yours. It wants you to understand what psychic skill you have. The goddess Hecate has the same intention for us, making this the perfect pairing.

This wise goddess knows there are many ways we can receive intuitive messages, from hearing or seeing them in our minds' eye, to sensing them in our bodies or seeing signs in the outside world.

HOW TO USE IT

Source a crystal you are drawn to. Labradorite can be easily faceted into different shapes, so look for one that is significant to you, such as the Hecate triple moon symbol.

Start by sleeping with the crystal next to you so it can enhance and develop your psychic skills while you sleep. Over time, more intuitive messages will start coming through.

After a while, you might like to use your crystal to communicate with Hecate directly. Look to her when you reach a crossroads and need help understanding what to do next. When this happens, list down all the options. Then, on a dark moon, light a black candle to invite in Hecate and her guidance. Hold your Labradorite in your dominant hand. Take some breaths and start considering each option. It might take some practice but, over time, good options will feel expansive and energizing, as if your body almost leans into them. A negative answer will feel depleting and sad, as though you are physically stepping away from the idea.

Additional properties: Grounding, protecting and ancestral, goddess connections

Sources: Madagascar, Canada, Finland, China and the USA

Crystal shapes: Shapes and spheres

Best times to work with this crystal: Monday, evening

OCTOBER CRYSTAL OF THE MONTH: BLOODSTONE

CONNECT WITH YOUR ANCESTORS

This month's energy clearly has us looking to the past, considering our relationships and our ancestors. There are two crystals that focus on our family tree, the first being Petrified Wood, which, as its name suggests, symbolizes the 'family tree'. Then there is Bloodstone, also called Heliotrope, which helps us connect to our bloodline.

There are many different past-life crystals, but this one helps us to tap into our DNA and brings forward the ancient wisdom of our ancestors, helping us to connect to them so we can ask for guidance.

CONNECTING TO YOUR ANCESTORS WITH BLOODSTONE

1. Hold your crystal in your dominant hand, close your eyes and focus on your breath. Visualize a pink circle of light around you. When you feel relaxed, welcome your ancestors and see them in your mind's eye standing by your side.

2. On a slow inhale, ask them a simple question, for example, how are they connected to you? Then let the answer come on a slow exhale.

3. Following the breath again, on the next inhale, ask another question, then allow the answer to come on the exhale.

4. Keep doing this until you have asked all of your questions, or until the messages stop. Thank your ancestors, let them step back, and then open your eyes.

Note that it will always be easier to connect to ancestors we don't know compared to those we do, as there won't be any expectations involved.

Additional properties: Samhain celebrations, past-life regression, enhances a family connection

Sources: India, the USA, Brazil and China

Crystal shapes: Hearts, tumbles and jewellery

Best times to work with this crystal: Monday, evening

OCTOBER ACTIVITIES

1. SAMHAIN TRADITION

If you like the idea of using Samhain to honour your ancestors, then each year you might consider offering your gratitude to them. This can simply be to show them that you care, but it can also help welcome them to come closer so they can better impart their wisdom in the form of intuitive messages and synchronicities

SAMHAIN HERBAL BLESSING

On page 116 I introduced the idea of herbal blessings: the burning of crystal-infused herbs, resins, leaves and petals to give thanks or ask for help. Following the steps previously set out, use these ingredients: rosemary (remembrance), rose buds (love); myth (soul connection), and Bloodstone crystal chips.

If you also want to seek guidance from them, add sage (connection and clearing), rose (ancestral connection) or juniper berries or mugwort (psychic connection).

Many people create a small Samhain shrine to honour past family members. Normally this includes photos, mementos of them and a candle to direct them. You might like to present it on a petrified wood slab, and this would also be the perfect place to make your offering.

2. CRYSTAL PROTECTION

Did you know that people, events or situations can drain us, leaving us feeling lethargic and even under the weather, especially if we are very empathic? This happens when they unknowingly sap our energy.

If you feel this could be happening to you, you can use Black Obsidian to protect your energies. Source a Black Obsidian crystal you can hold and do this each morning before you leave your house:

CRYSTAL ENERGY PROTECTION

This is a quick practice that only takes 30 seconds.

1. Hold your crystal in your dominant hand.
2. Stand up and take some deep breaths.
3. Imagine a circle of light around you – this is your auric field. Around that, visualize a thin black line appearing, and have the intention that it represents the crystal's energy protecting you.
4. Repeat the process every time you know you are going into a difficult situation.

ROSE QUARTZ PROTECTION ENERGIES

When you are going into a difficult situation, I suggest adding a black crystal line around your aura. The downside, however, is that other people may interpret it as 'closed-off' or negative body language. Therefore, when you are in place with people you know won't drain your energy, you could wear a Rose Quartz crystal and see a pink line surrounding you instead. Rose Quartz will create self-caring, protective energy for you, but those around you will read your energies as open and caring.

November

AUTUMN ENDS: **SELF DEVELOPMENT, AMBITION & SELF CARE**

By now the home fires are lit, our favourite jumpers are an everyday essential again, hot foods help keep us warm and we might need a scarf and gloves when we head outside. This is your last chance to do the jobs that would prove impossible or too complicated in winter.

November is still a Water element month, so emotions feed our behaviour, but as the Air element of winter starts taking over, we move from the deep-seated Scorpio emotions into more heady emotions. Scorpio has created space for deep healing. Sagittarius looks ahead, asking what is possible in the coming year and who we want to be now that those changes have been made.

November is an opportunity to pause before December's festive celebrations begin. It is the time to contemplate and enjoy Sagittarius' passion for philosophical ideas, new inspirations and potential new goals. If you find yourself daydreaming over your Thanksgiving meal, this might be why.

This month I will show you how to use crystals to connect to the ancient Greek goddess Iris and practice temperance, balance your emotions and treat yourself to some much-needed self-care.

November facts
Colours: Pink, purple and light blue
Crystal shapes: Water droplets, towers and flames
Seasonal chakra energies: Root and crown chakras
Animals: Dogs, beavers and dolphins
Seasonal shrine/altar: Vision board, pink candles
Intention setting/affirmation: *I honour and look after myself*
November birthstone (traditional): Imperial Topaz
Birthstone properties: Assertiveness
November birthstone (modern): Citrine
Birthstone properties: Pure joy and happiness

NOVEMBER ELEMENT:
AUTUMN & WATER

The nights draw in, and we focus on our internal world. At this time of year, we look at who we are and cultivate our own opinions on different aspects of our life, as well as life itself. These are moments of true, deep contemplation, with a touch of philosophical thinking. This work is motivated by the season's element – Water – which controls our emotions.

Oceans cover 71 per cent of the Earth's surface and 96 per cent of the water on Earth is found in our oceans; where life once originated. There is also water underground and in all plants and animals, including ourselves.

Like crystals, water can be programmed with information. The human body is made up of between 50–75 per cent water, and this water can be programmed as well. When we think negative thoughts, the water in our bodies can become programmed with that thought. This is another reason why cultivating a positive mindset and raising our vibration is so important (see page 42). It is also why I love crystal elixirs because they allow us to consciously consume water with a positive vibration.

We can also use water to cleanse ourselves; going for a swim or taking a bath or showering can wash away bad energies. This is a really useful cleansing method if you feel drawn to oceans or enjoy the sound of running water.

Water is said to be a divine feminine element. It is also connected to the moon, controlling the ocean's tides and heightening our intuition. It can change its physical form, connecting to the Fire and Air elements as it becomes vapour and water droplets, or leaning into the Earth element as it forms into ice. In this form it magically becomes a crystal.

CRYSTALS FOR AUTUMN'S WATER ELEMENT

Emotional support: Ocean Jasper

Releasing old emotions: Aquamarine

Anxiety: Larimar

NOVEMBER ASTROLOGY
& FULL MOON

♏♐

We start November with Scorpio asking us to focus on those big emotions that might be affecting our lives in obvious as well as unexpected ways. Remember that as we travel through an astrological month, the energy of each season heightens and then subsides so even if emotions feel complicated right now, they will calm.

On 23 November, our attention shifts to Sagittarius season, which asks us to contemplate what goals we want to achieve in the coming year, and therefore what self-development we need to do now to make those goals happen. In essence, we go from healing the past, to identifying blocks that might stop us achieving future goals.

Sagittarius is the last full astrological season of the year, and it wants us to look forward and see what is coming. Its symbol, the archer and their bow, indicates Sagittarius' ability to look ahead and pinpoint where it wants that arrow to fall next.

Why is Sagittarius so forward-focused at a time of year when Mother Nature and our energies want us to relax and stay indoors? The answer is that the zodiac knows we need an injection of positivity after the deep emotional pools that Scorpio season may have had us swimming through. It knows that the darkest months of the year are not the time to be exploring our most complex emotions, so Sagittarius steps in with a sense of adventure and increased optimism to help us shake off the deep work we have been doing. It allows us to go into winter with emotional reserves, and dreams of the coming year.

NOVEMBER'S FULL MOON

This moon is named Beaver Moon by Indigenous Americans because at this time of year beavers strengthen and prepare their dams in preparation for the cold winter months, much like us.

NOVEMBER GODDESS: IRIS

NAME AND MEANING

Daughter of the sea god Thaumas and water nymph Electa, the Greek goddess Iris, whose name means 'rainbow', was a servant to queen Hera, wife of Zeus. Iris was one of her many messengers, and would travel down from Mount Olympus using her rainbows to send messages to the mortals. It is not surprising, then, that she is the goddess of rainbows and communication.

CHARACTERISTICS

Married to Zephyrus, the god of the west winds, Iris gave birth to Eros, the god of desire. There is little known about this intrinsically uplifting goddess. Images of Iris normally show her with wings, which speak of her mother's gifts, or holding a jug of water, honouring her father. Her affinity with the Water element and the Air element connects her to Sagittarius, which is also connected to both those elements. Its astrological season starts in autumn, the Water element season, then transitions to winter and the Air energies it offers.

TEACHING

The rainbow is also the perfect metaphor for Sagittarius' aspirations – when it looks to the sky to fire its arrow, it is aiming for the pot of gold at the end of the rainbow. In tarot, the Temperance card is related to Sagittarius. In the famous Rider-Waite-Smith tarot deck, Iris is thought to be shown on this card, depicted as a winged being pouring water between two cups. She has one foot dipped in a pool of water, with iris flowers in the grass behind.

She is inviting us to dip a toe into future ideas, but also wants to remind us not to move too quickly, a behaviour that Sagittarius can fall victim to. You might want to rush ahead and start planning for the coming year, but November is also a time to practise stillness and temperance.

SYMBOLS OF IRIS

Rainbows, the iris flower, caduceus and pitcher

NOVEMBER WATER ELEMENT CRYSTAL:
OCEAN JASPER

BALANCING EMOTIONS

Sometimes the name of a crystal is the biggest clue to its properties – and this is one of those times. The name Ocean Jasper indicates that we are dealing with a Water element crystal that has the ability to deal with vast oceans of emotions. It is sometimes called Orbicular Jasper.

This crystal can go to the deep depths of our worst traumas, raising them to the surface so they can be released and healed. It can also show us where our emotions ebb and flow and help to stabilize them.

Ocean Jasper can look like an abstract painting of different colours and tones mixed together, which means it can support different aspects of our lives. It is available in every colour imaginable, so when choosing a specimen it is a good idea to find one that has the colours to match your needs (see page 21 for help with this).

HOW TO USE IT

The ebbs and flow of the tides are controlled by the moon, which also represents our emotional body. Therefore, it is a good idea to do deep emotional work around a full moon so we can elicit its support.

Ocean Jasper moon water combines the energy of the moon with your crystal to help calm stormy waters and bring emotions into balance. On page 106 I explained how to make a solar elixir. Follow the same process, but this time use an Ocean Jasper crystal and put your water outside overnight when there is a full moon, so it can be charged with its energy.

Additional properties: Balance, understanding and moon ceremonies

Source: Madagascar

Crystal shapes: Spheres, tumbles and water droplets

Best times to work with this crystal: Monday or Saturday, evenings

NOVEMBER ASTROLOGICAL CRYSTAL: SAMADHI QUARTZ

ENLIGHTENMENT

Sagittarius is about pinpointing who we want to be and where we want to go, and then creating a plan to get there. Sometimes it can be a little too focused for some people: when its mind is set, no one can stop Sagittarius energy.

All Fire signs like to focus on passions and goals. Sagittarius season starts in autumn, the Water element season, and ends in Air element winter. When combined, this indicates a time of year when we go through emotional healings, then on to realizations that start our spiritual awakening, with the ultimate goal of one day reaching enlightenment: moments when we are fully connected back to universal consciousness and a state of pure joy.

The higher the geographical location in which a Clear Quartz is found, the higher its vibration will be. Therefore, crystals found in the Himalayas have the highest vibration of all. Samadhi Quartz is the name given to the crystals found at the highest possible elevation. Its name literally means 'enlightenment', and these crystals are found at elevations so high that it affects their mineral composition, creating a beautiful Pink Quartz crystal. Perhaps this is an indication that we reach enlightenment by simply returning to the heart.

HOW TO USE IT

Find a statement piece that can be displayed somewhere prominent so your crystal can fill your home with its energy and make slow, incremental changes to your life. Then, during the Sagittarius season when you're actively working on your self-development or other big goals, bring the crystal into your workspace so it can support the work you are doing.

Additional properties: Compassion, spiritual insights and joy

Source: The Himalayas

Crystal shape: Natural clusters

Best times to work with this crystal: Sunday, all day

NOVEMBER GODDESS CRYSTAL:
SCOLECITE

PEACE

Iris steps in this month to ask us to practise temperance. Even though Sagittarius energies want us to start new projects right away, Iris reminds us that we are heading into winter. Yes, we can dream and put things in place for future goals, but it is also a time for self-development, self-care and moments of silence, because in those moments of pure silence, true divine guidance can come through.

Scolecite holds the energy of pure peace, the vibration of the first sound: *Om*. In its raw form, its striking appearance is made up of numerous needle points coming out of all sides, sending its energy of peace to all areas of a room.

It is also possible to find this crystal faceted into different shapes. Palms are perfect for meditation, while necklaces and rings can help keep you calm and centred when you might otherwise feel stressed.

HOW TO USE IT

Your energetic, emotional and physical body will all benefit from taking moments to relax and concentrate on being at peace. We need to focus on the energy body if we also want to support our emotional and physical body. For this season, I recommend meditating in a circle of small Scolecite tumbles or chips.

These energies are so subtle that you could place a tumble on the four sides of your bed and sleep within its calming energy.

Additional properties: Deepens meditations, calms the functions of the body, creates a healing space

Sources: India, Iceland and Australia

Crystal shapes: Tumbles, chips, hearts and palms

Best times to work with this crystal: Sunday to Monday, dawn

NOVEMBER CRYSTAL OF THE MONTH: ROSE QUARTZ

LOVE AND SELF-CARE

Our emotions might feel pushed and pulled this month, as Scorpio asks us to embark on some deep healing – then Sagittarius comes in asking us to consider new projects, while the goddess Iris reminds us to step back.

The nights are growing longer and we want more sleep. The vitality of the summer months has left us, and we seek reserves for December's festivities. Everything is telling us November needs to be a time for self-care. Rose Quartz is therefore the perfect crystal to keep with us all month, carrying it everywhere as a talisman as a reminder to check in with yourself to see how you are, and what your emotional and physical body needs.

Regularly cited as the crystal of love, Rose (or Pink) Quartz is that and so much more. It holds the answers to the timeless statement: *You cannot truly love someone else until you learn to love yourself.*

But what does that mean? The answer is different for everyone, but perhaps it involves coming to terms with parts of ourselves we are not happy with. Devoting time for our own wellbeing and releasing ourselves from feelings of guilt and the misconception that occasional self-care is selfish.

HOW TO USE IT

In my experience, if someone hasn't done much self-care work before, they sometimes find working with Rose Quartz quite uncomfortable. How uncomfortable it is can be an indication of just how much self-care work is required. Therefore, I'm going to suggest three different activities, each incrementally increasing the amount of time spent with Rose Quartz and its self-care energies. Choose the one you feel is best for you.

1. Display a Rose Quartz specimen in your living space.

2. Make a Rose Quartz crystal elixir (see page 46). Start with one sip a day, inceasing the sips each day until you can drink the full glass.

3. For a self-care Rose Quartz bath, follow the steps on page 79.

Additional properties: Romance, inner beauty and supports the heart chakra

Sources: Madagascar, Brazil and China

Crystal shapes: Raw, spheres, hearts, towers and tumbles

Best times to work with this crystal: Friday, evenings

1. DEEP SLEEP SELF-CARE RITUAL

One way we can be guided by November's request to love ourselves more is to make sure we get enough sleep, giving our brain and body time to repair and go through its detoxing process.

Below is a recipe for an essential oil roller that you can infuse with three different crystals: one is a general self-care sleep time recipe; the other two support deep sleep and help those who struggle with sleep problems.

CRYSTAL-INFUSED ESSENTIAL OIL ROLLER

To create your roller, you first need to make an indirect crystal elixir in the morning sun. To do this, follow the guidance on page 46 and use your choice of one of these crystals:

For problem sleepers: Celestite

For deep sleep: Howlite or Scolecite

For a self-care sleep routine: Rose Quartz or Amethyst

To make your essential oil roller, you'll need a pipette and a small funnel; your crystal elixir; a brown glass roller bottle; some Amethyst crystal chips; a stabilizer, such as vodka or vegetable glycerin; organic jojoba or coconut oil; and one of the following essential oils:

For problem sleepers: Bergamot or lavender oil

For deep sleep: Vetiver or cedar wood oil

For a self-care sleep routine: Lavender or cedar wood oil

To make it, use a pipette to put 2 drops of crystal elixir and 3–5 drops of essential oil into the bottle. Add a teaspoon of crystal chips. Then use the funnel to add 1 tablespoon stabilizer and fill the rest of the bottle with your organic oil. Add the roller top and shake well before use.

2. LAVA STONE & ESSENTIAL OILS FOR SELF-DEVELOPMENT

Self-development is another big topic this month. As you dream of potential achievements for the year to come, consider what aspects you might find hard and what self-help work you could do to support it. Use these journal prompts to help you understand where to put your focus:

1. Think back to your last goal or the last project you worked on. What aspect of it did you find hard? Can you remember if you experienced any negative mind chatter? If yes, what came up?

2. How could you support your physical health more?

3. Now consider things you can implement into your life. For example, would you like to try any of the following: morning routines, meditation, gratitude practices, improving your intuitive abilities or reading more self-development books?

Your answers may help you see what needs work. To use crystals for healing and making positive changes, I recommend using Lava Stone with essential oils. Lava Stone is rock formed from a volcanic eruption. It is highly porous and is often made into beads for jewellery. Essential oil can then be dropped on the beads, which sucks it in, allowing us to carry your essential oils with you.

Below is a list of self-development areas and suggested essential oils that will help. Use a lava stone bracelet so you can keep that oil with you, and it can motivate, even heal you:

Loneliness
Cedarwood

Emotional release
Geranium

Focus on gratitude
Peppermint

Self-confidence
Lavender

Anxiety
Bergamot

Living in the moment
Patchouli

Procrastination
Lemon

Intuition development
Clary sage

Heartache
Rose

Meditation
Sandalwood

December

WINTER BEGINS: **ABUNDANCE, CELEBRATION & MANIFESTATION**

For many, the festive season begins with opening the first window of their advent calendar. For others, Hanukkah sees a candle lit on the Menorah each night. Nowadays, this is the month of work parties, followed by reunions with family and friends. The excitement peaks around Yule, with the traditional Christmas celebrations, before we collectively ring in the New Year.

Our ancestors kept rations aside for December's revelry, knowing how important celebrations were for morale, and they had something to celebrate, knowing lighter days would slowly return after the winter solstice.

We finish the year as we began: in the Air element season. December starts with Sagittarius still offering mixed messages, one day motivating us to focus on self-development and self-care, the next asking us to dream of what the new year holds. Capricorn takes over, holding space for us to rest and reflect. December ends with the New Year, a wave of celebration and global unity, creating the perfect environment for manifesting.

This month we use crystals to connect to the ancient British goddess Elen of the Ways to illuminate your path ahead. I also show you which crystals can help you reflect, cleanse and manifest an even more magical new year.

December facts
Colours: Red, green and white
Crystal shapes: Raw and skulls
Seasonal chakra energies: All the chakras
Animals: Reindeer, foxes and goats
Seasonal shrine/altar: Evergreens, holly, mistletoe, Christmas rose, red and green candles, Yule log, symbols of Christmas and the New Year
Intention setting/affirmation: *I manifest my ideal year*
December birthstone (traditional): Turquoise,
Birthstone properties: Wisdom
December birthstone (modern): Tanzanite
Birthstone properties: Diplomacy

DECEMBER SABBATH CELEBRATIONS: WINTER SOLSTICE, YULE & THE TWELVE DAYS OF YULE

The winter solstice occurs annually from 20–23 December. It marks the point at which the Earth's tilt is furthest away from the sun, meaning that the northern hemisphere experiences its shortest day. Major monuments like Newgrange in Ireland, Stonehenge in the UK and the Cahokia Woodhenge in the USA all mark the point of sunrise on this date. It is a moment to celebrate the return of the sun. In neo-pagan tradition this is the moment the Oak King, representing the light, retakes his throne from the Holly King, the dark.

Our ancestors may have been rationing their food for months, but today they would feast to honour how well they had done and raise their hopes for the future. Homes were decorated with holly and mistletoe. Both symbols of hope, they reminded people that life keeps going, even in the darkest times. Lighted candles were placed in the windows for protection, while the Yule log burned all night in the hearth, warming the home. Its ashes would be stored and used to bless future crops.

The word 'Yule' has possible ties to the Saxon word *hweol* ('wheel'), referring to the wheel of the year or the great wheel in the sky, the sun. This celebration is part of a longer ancient pagan holiday known as the Twelve Days of Yule, also called Yuletide. Yule starts at dusk the day before the winter solstice celebrations. Early Europeans saw these twelve days as a hidden year created by the gap of time between the lunar and solar calendar. Each day represented a whole month. It was a chance to look back and offer gratitude for the correlating month in the previous year and focus on manifesting abundance for that month in the year ahead.

CRYSTALS TO DISPLAY OVER YULE

Cultivate hope and joy: Crazy Lace Agate and Peridot

Manifest a successful year to come: Rutilated Quartz and Citrine

Giving thanks for the previous year: Mangano Calcite

DECEMBER ASTROLOGY
& FULL MOON

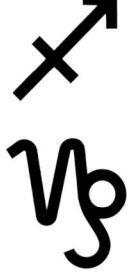

Sagittarius season joins us as we step into December, and this means we still feel it enticing us to dream of what the coming new year might bring. On 22 December we switched again from an energetic, jovial sign to the more serious, considered energies of Capricorn. This is the energy that inspires us to make New Year's resolutions. It invites us to do a life audit, an evaluation of where we are, where we are going and what might need to change in order for us to get there.

An Earth sign, its symbol is a capricornus: half mountain goat, half fish. If you feel aspects of your life shifting, perhaps even falling away this month, that's the mountain goat testing where there is instability in your life. Strong foundations is a theme that appears several times throughout the astrological year, as if the universe is constantly asking us to check if our plans are built on solid ground.

This time, future-focused Capricorn is helping us let go of the old, so that something better can be created in the new year. We are asked where in our lives are we playing small and what needs to change for us to realize our purpose. And, with Saturn being Capricorn's planet, we might feel a dramatic push from that mountain goat as the universe helps us realize our full potential.

DECEMBER'S FULL MOON

Most commonly, this moon is known as the Cold Moon. However, it can also be called the Moon Before Yule or the Oak Moon. Both honour the winter solstice, and the Oak Moon in particular marks the return of the Oak King (the sun).

DECEMBER GODDESS: ELEN OF THE WAYS

NAME AND MEANING

Elen of the Ways is our final goddess. She is elusive and changeable, with references to her spanning all the way back to the Palaeolithic Age. She is the goddess of sovereignty, roads, paths and ley lines, and her name means 'torch'. The most famous account of her is 'The Dream of Macsen Wledig', from a collection of 12th–13th-century Welsh traditional stories called the Mabinogion.

In it, Macsen Wledig, most likely the Roman Emperor Maximus, dreamed he travelled and found Elen, falling in love with her. When he woke, Macsen sent his men to find her, but when they did Elen said Macsen must visit her if he wanted to marry her – so he did. Macsen was so in love that he made Elen Empress of Rome. For her wedding gift he gave her Britain for her father to rule. He also granted her wish that three castles should be built to defend Britain, along with roads to link them all, hence the name Elen of the Ways.

CHARACTERISTICS

Elen is regularly depicted wearing antlers, indicating just how important deer were in the Palaeolithic Age. It also honours the strength of female willpower and symbolizes female reindeer, who intuitively lead their herds annually on mass migrations. Elen is often shown walking with them, lamp in hand.

TEACHING

The Roman roads created in her name are symbolic of how Elen can support us today: she helps us find the best way forward and decide what path we should take through life. This festive goddess can guide us through Capricorn season, helping us make new paths that will have long-lasting outcomes. She stays with us as we say goodbye to the current year and start navigating what we want the coming new year ahead to be.

SYMBOLS OF ELEN OF THE WAYS

Reindeer, emerald green, lanterns, paths and ley lines

DECEMBER SABBATH CRYSTALS:
TWELVE DAYS OF YULE

The Twelve Days of Yule creates a magical space in our calendar to celebrate the past year and send positive energies to the year ahead. This is the perfect time to journal about the previous year and bring to rest anything you want to leave behind, while also using the magical energies our ancestors associated with this time to start manifesting the year ahead.

If you're not sure what you want to manifest, consider what goals Sagittarius season had you dreaming of, as well as the areas Capricorn season is highlighting that may need stronger foundations.

There are multiple legends for each day of Yule, and over time, themes started to appear. Below is a table depicting each day of Yuletide, its theme, which correlates with the theme of its month, and the crystal that will support it.

The dates relating to Yule change slightly each year, but there is a page on my website gemmapetherbridge.com/sabbaths with the current dates. Head there to see when Yule begins this year.

TWELVE DAYS OF YULE END-OF-YEAR RITUAL

Each of the twelve days represents one month from the previous twelve months. Allow yourself 30 minutes to work on the journal prompts below, followed by the manifestation visualization. Keep the relevant crystal companion near as you complete these journal prompts:

1. Reflect on that month in the previous twelve months. What did you enjoy? What emotions do you want to release?

2. Now consider what that day's theme means to you? Can you do something today and in that coming month that honours that theme?

3. What would you like to happen during that month in the coming 12 months?

For the manifestation visualization, hold your crystal companion in your dominant hand and visualize how you want that month to play out in the year to come.

DAY AND CELEBRATION	MONTH CELEBRATED	THEME AND EVENTS	CRYSTAL COMPANION
Day 1: Mother's Night Celebrations (a time to honour the divine feminine)	January	Time: Honouring the year we are leaving behind, and the year ahead	Peach Moonstone (see page 84)
Day 2: Winter solstice	February	Hope: February marks winter drawing to an end	Peridot (see page 102)
Day 3	March	Courage: Our ancestors had made it through winter	Red Jasper or Tiger's Eye
Day 4	April	Love: Newborn animals all around	Pink Kunzite
Day 5: Normally Christmas Eve is today	May	Community: May's community celebrations combines with people coming together on Christmas Eve	Yellow Muscovite or Mica
Day 6: Normally Christmas Day is today	June	Abundance: Nature is in full growth. On Christmas Day abundance is experienced as gifts and plentiful meals	Rutilated Quartz (see page 140)
Day 7: Normally Boxing Day is today	July	Fun: The joy of being outside in the summer sun and celebrating with loved ones on Boxing Day	Citrine (see page 118)
Day 8	August	Clarity: The bright summer sun and Lion's Gate in August	Clear Quartz (see page 188)
Day 9	September	Creativity: Harvesting the fields and preparing for winter	Orange Calcite (see page 48)
Day 10	October	Self Reflection: Samhain and Scorpio season. A time for deep healing	Black Moonstone (see page 34)
Day 11	November	Energy: A time for self-care	Rose Quartz (see page 174)
Day 12: New Year's Eve	December	Wisdom: Insights the twelve days have offered you	Lapis Lazuli (see page 76)

DECEMBER ASTROLOGICAL CRYSTAL: BLACK TOURMALINE

GROUNDING

The mountain goat can famously navigate terrain most of us cannot climb, and this means Capricorn can see aspects of our lives that haven't been built on strong foundations.

Those foundations might wobble this month, and by doing so are highlighted to us. Once they are highlighted, you have the free will to make changes. In reality even small changes, like considering different perspectives on situations, can firm up those foundations. Nevertheless, when we feel that shift under our feet it can bring up anxieties.

Black Tourmaline (also called Schorl) is a root-chakra crystal that can ground our energies, even in stressful situations. Anxieties bring a lot of energy into the head. Black Tourmaline counteracts this by moving the energies back down through the body so we calm ourselves and make better decisions.

HOW TO USE IT

Lie down and place a Black Tourmaline crystal under your feet. Stay there for as long as you can. You might want to put some meditation music on, do a guided visualization or take a nap. This gives the body a chance to relieve any stress and start to relax.

Additional properties: Protection, safety and stability

Sources: Brazil, Africa, Pakistan and the USA

Crystal shapes: Raw, palms and jewellery

Best times to work with this crystal: Any day, anytime

DECEMBER GODDESS CRYSTAL:
SMOKY QUARTZ

GUIDANCE

When you don't know what path to take, Smoky Quartz (sometimes called Morion Quartz) is there to light the way for you. Museum-quality specimens show a dark Smoky Quartz transitioning through the spectrum to a lighter shade. This also represents the main properties of this crystal, as it takes you from the dark (the unknown) to the light (the known).

I personally see this as our spirit guide crystal. Just like Elen of the Ways, when we are standing at our own metaphorical crossroads and we don't know which route to take, it wants to help us find the answers.

HOW TO USE IT

When you have a decision to make and you don't know which option is best, or you are coming to the end of a phase in your life (including the end of the year) and you want to start exploring your options, this is a good time to start wearing Smoky Quartz jewellery.

Wearing this crystal as jewellery gives it time to work with us, bringing to our attention the information we need to understand what our next steps might be. You might want to consider a ring, so that it's positioned next to its correlating chakra, the root chakra. Consider wearing your Smoky Quartz jewellery throughout Sagittarius season as well, so it can help you to refine all of the ideas that come through.

Additional properties: Planning, intuitive guidance and grounding

Source: Worldwide

Crystal shapes: Raw, points and jewellery

Best times to work with this crystal: Sunday, late afternoons

DECEMBER CRYSTAL OF THE MONTH: CLEAR QUARTZ

CLEANSING

This is the first crystal many people collect. Clear Quartz (also called Rock Quartz) is the most common crystal in the ground, so it's an easy one for us to start working with as its energies feel more familiar. Clear Quartz can work with all our chakras, and can be programmed with any task (see page 17).

It can also amplify the energies of other things. For example, if you display Clear Quartz next to other crystals it heightens their energy, or you could wear it when you're celebrating (consider the manifesting energies of New Year's Eve) so it can heighten your enjoyment even more.

Finally, it can cleanse and release unwanted energies from other crystals, spaces and even from ourselves. This makes it the perfect crystal tool to support you through all the themes the Twelve Days of Yule covers.

HOW TO USE IT

If you want to clear old emotions but you don't know which crystal correlates with those emotions, you can use Clear Quartz because of its ability to support all areas of our lives.

If you are planning to do the Twelve Days of Yule end-of-year ritual on page 182, you can use two Clear Quartz tumbles or a pair of harmonizing crystals to cleanse any emotions you don't want to take into the next year.

Each day, once you have completed the journalling prompts and identified any emotions you want to cleanse, hold one Clear Quartz crystal in each hand. Think of the event and the emotions involved, then have the intention that the emotion is going into the crystal to be cleansed. It's as easy as that!

Additional properties: Amplifies, cleanses and programmes

Source: Worldwide

Crystal shapes: Points, palms, skulls and sacred geometry sets

Best times to work with this crystal: Sunday, early afternoons

DECEMBER ACTIVITIES

1. LITHOMANCY

Our ancestors believed each day of Yule held clues to how the twelve months ahead would unfold. Their emotions, the events of the day and even the weather were all seen as signs, and people often used divination to gain insight. Crystals can be used in the same way you would work with oracle cards. This ancient art, known as lithomancy, involves choosing crystals and 'reading' their properties to receive guidance. During each day of Yule, you might like to choose a crystal that will represent the energies and theme of that month in the coming year. This is what you do:

USING CRYSTALS AS ORACLE CARDS

Source at least seven crystal tumbles that have different themes/properties and cover different topics. These are the seven I usually suggest starting with:

1. Rose Quartz: Love or self-care
2. Black Tourmaline: Protection
3. Clear Quartz: Spirituality and clarity
4. Selenite or Moonstone: The start, middle or end of something
5. Emerald or Green Aventurine: Learnings or growth
6. Citrine: Goals or career
7. Angelite or Celestite: You are being supported/guided

Put your tumbles in front of you, close your eyes and relax. Have the intention that you're choosing a crystal whose properties match the month in the coming year you are focused on. Hover your hands over all the crystals and choose one. The properties of the crystal you have selected will give you some idea of the themes you might experience at that time next year.

Once you know how to do a standard reading, elaborate on it by including the features on the crystals to add detail. For example, cracks in a crystal might suggest problems, while a rainbow could be an amplification of that theme. Trust that you will notice a feature only when it's relevant for that reading.

2. CRYSTAL SKULLS

We finish the year by exploring and working with crystal skulls. Examples of them have been found in ancient civilizations, but we don't know the reason for their creation, or when they first appeared. One suggestion is that it's easier to programme information into a crystal when we can understand where we are sending it (the brain). Another theory is that it's easier to relate to a crystal if it looks like us because it makes the crystal more personable and makes us more susceptible to receiving intuitive messages from it.

As the year ends, I invite you to look at the answers in your Twelve Days of Yule ritual. Ask yourself: *Who do I want to be this time next year?* Now look for a crystal skull you are drawn to. Many people start with Clear Quartz skulls because they are easier to programme. However, as this one is going to be your talisman for the year ahead, to remind you of your long-term goals, I recommend choosing a specimen you like and will enjoy seeing every day.

NEW YEAR'S EVE CRYSTAL SKULL CIRCLE

New Year's Eve is the best time to programme your crystal skull to make the most of its positive manifesting energies. Prepare your space by cleaning and clearing things away, then add some candles, essential oils and music that lifts your spirits. Programme your crystal skull by following these steps.

1. With your crystal in front of you, sit down and calm your mind. Place your non-dominant hand on the crystal.

2. Imagine who you want to be this time next year, where you are, what you are doing, who you are with and what you have achieved and experienced. Ask your future self how it feels to have done all of that.

3. Experience the emotions, intentionally sending them into the crystal. After a few minutes, take your hands away from your crystal and open your eyes.

4. Display your crystal in a place where you will see it all year, so it can attract those experiences into your life. Then go and enjoy the New Year celebrations!

About the Author

Gemma Petherbridge set out on her spiritual path as a small child, prophesizing in her dreams. Seeing spirits, naturally intuiting situations and gaining insight into the destinies of others came as second nature, and after losing her parents at a young age she turned to spirituality for answers. Her journey into the world of wellness and holistic therapies began in earnest when, aged 23, she studied hypnotherapy.

Seventeen years on, Gemma is a Certified Crystal Healer, Intuition Teacher and Akashic Records Reader. Having transitioned from healer to teacher, she has now taught and inspired thousands of people worldwide. In 2017 she founded Conscience Crystals, which offers workshops, courses and an online shop. With her growing following, Gemma is now regularly asked to lead workshops and speak at holistic events and festivals, and major international businesses seek her guidance in incorporating crystals into office environments and products. She also presents the spiritual and holistic wellbeing podcast Crystal Mystery School. Her first crystal book, *The Crystal Apothecary* launched worldwide in 2022, with her Crystal Mystery School launching in 2024.

Instagram: @gemmapetherbridge
Website: gemmapetherbridge.com

Acknowledgements

Thanks to Holly Booth, Sarah Ann Wright and Michelle Szpak for their photos.

To Julie Evans of Nourish To Bloom for your essential oils pearls of wisdom, and Melanie Smith of Orkney Mangata for her candle-making knowledge. And for herbal advice, thank you to Rachel Morley, founder of Embodied Energy Training and Rachel Drysdale of Fire + Alchemy, London.

I am blessed to be supported by a publishing house that is a joy to work with. Particular gratitude goes to Nicky, Louisa, Leanne, Yasia, Claire, Giulia, Jen, Allison, Charlotte and Ailie.

Love Gem